W0114626

ON THE DIRECT NUMERICAL
CALCULATION OF ELLIPTIC
FUNCTIONS AND INTEGRALS

ON THE DIRECT NUMERICAL CALCULATION OF ELLIPTIC FUNCTIONS AND INTEGRALS

By

LOUIS V. KING, M.A., D.Sc., F.R.S.

Macdonald Professor of Physics,
McGill University, Montreal

CAMBRIDGE
AT THE UNIVERSITY PRESS
1924

DEDICATED TO THE MEMORY OF
JAMES HARKNESS
Peter Redpath Professor
of Pure Mathematics,
McGill University
1903–1923

CAMBRIDGE
UNIVERSITY PRESS

University Printing House, Cambridge CB2 8BS, United Kingdom

Cambridge University Press is part of the University of Cambridge.

It furthers the University's mission by disseminating knowledge in the pursuit of education, learning and research at the highest international levels of excellence.

www.cambridge.org
Information on this title: www.cambridge.org/9781316612729

© Cambridge University Press 1924

This publication is in copyright. Subject to statutory exception and to the provisions of relevant collective licensing agreements, no reproduction of any part may take place without the written permission of Cambridge University Press.

First published 1924
First paperback edition 2016

A catalogue record for this publication is available from the British Library

ISBN 978-1-316-61272-9 Paperback

Cambridge University Press has no responsibility for the persistence or accuracy of URLs for external or third-party internet websites referred to in this publication, and does not guarantee that any content on such websites is, or will remain, accurate or appropriate.

PREFACE

WHILE lecturing in 1913 on the calculation of self and mutual induction, the writer noticed that the integral in the case of co-axial circles was of such a form as to make the application of the arithmetico-geometrical means obvious. An extremely elegant formula (*Proc. Roy. Soc.* A, Vol. c. 1921, p. 63. Appendix, this book, No. 35) when put to numerical test justified the method as one well adapted to computation. The calculation of more complex cases of self and mutual induction led to several interesting formulae for the complete third elliptic integral. With a view to making a complete study of the group of recurrence formulae associated with the A.G.M. scales, further work was postponed until a vacation in 1921, when many of the results included in this book were obtained, and were thought by the writer to be new. On later consulting the *Collected Works* of Legendre, Gauss and Jacobi, the writer found that he had rediscovered many known formulae, but that quite a number, likely to be of service in computation, appeared to be hitherto unknown. In the summer of 1922 the writer decided, as a vacation task, to include in a single monograph the entire theory of elliptic functions associated with the A.G.M. scales, thus completing, and to some extent adding to, the work of Gauss left unpublished after his death, and placing before mathematicians in accessible form a mode of approach to Elliptic Function Theory directly related to the art of machine-computation. It is interesting to note in this connection that this subject was to have formed the content of the third volume of Halphen's *Fonctions Elliptiques*, unfortunately left incomplete at the time of his death.

While the present monograph contains many new formulae and methods of computation, no claim is made as to novelty in fundamental analytical treatment. References to more important books and memoirs consulted by the author are given in foot-notes. A complete bibliography is not given since the reader will find exhaustive references in the *Royal Society Index*, 1800–1900, Vol. I, Mathematics, and after 1900 in the mathematical volumes of the *International Catalogue of Scientific Literature*.

The author desires to express his thanks to his late mathematical colleague, Professor James Harkness of McGill University, for reading over the manuscript and proofs, and also to Mr Arthur Berry of King's College, Cambridge, for revising proofs, verifying formulae and for calling the writer's attention to several slips of calculation in the main text and appendix.

Finally, the author wishes to thank his colleague, Dr A. S. Eve, F.R.S., for his aid in obtaining from McGill University that financial assistance which has made possible the publication of this little volume. For the courtesy of the officers of the Cambridge University Press in undertaking to print this book in its usual impeccable style, the author is also deeply grateful. For assistance in the final revision of the proofs, the author is indebted to one of his students, Mr W. L. Robertson.

L. V. K.

McGill University,
April 18, 1924.

CONTENTS

ON THE DIRECT NUMERICAL CALCULATION
OF ELLIPTIC FUNCTIONS AND INTEGRALS

SECTION I

INTRODUCTION

ALTHOUGH the numerical evaluation of elliptic functions and integrals is dealt with in many of the standard treatises, it cannot be said that the use of tables is entirely satisfactory in dealing with computations required in several branches of physics and astronomy. Many of the integrals appearing in these problems can only be expressed as somewhat complicated expressions involving the complete elliptic integrals of the first and second kinds, and the corresponding functions of complementary modulus. These formulae give rise in many cases to differences of nearly equal quantities: the numerical calculation of such expressions to a sufficient number of significant figures is thus often a matter of some difficulty, requiring tedious and sometimes uncertain interpolations.

While the numerical calculation of the Theta-functions is extremely rapid by the use of the highly convergent series involving the *nome* q, these are not usually the functions which make their appearance directly in physical or astronomical problems. The necessary transformations required to express the results in terms of the q-series require in many cases somewhat complicated analysis, while numerical computation by this method necessitates the exact calculation of a large number of auxiliary quantities*.

The forthcoming tables in course of construction under the auspices of the British Association† are so arranged as to allow a large number of functions to be evaluated, and will very greatly aid numerical work in many practical problems involving the use of elliptic integrals and functions. While sufficient for most purposes, the use of these tables will still involve considerable interpolation when it is required to take out values corresponding to arguments and moduli expressed by six or

---

* On this point see Rosa, E. H., and Grover, F. N., *Bulletin* 169, Bureau of Standards, Washington (1913), pp. 67 and 73; Nagaoka, H., *Tok. Coll. Sc. J.* 27 (1909), No. 6, and *Jour. Tok. Math. Phys. Soc.* 4 (1908), p. 284; 6 (1911), p. 10; Nagaoka and Sakurai, *Scientific Papers of the Institute of Physical and Chemical Research*, Vol. II. Dec. 1922, pp. 1–67; Olshausen, G. R., *Phys. Rev.* 51 (Dec. 1910), pp. 617–636.

† See report (by Sir A. G. Greenhill), *Brit. Ass. Report*, 1912 (Dundee), pp. 39–55.

seven significant figures, as is now necessary in several modern applications of elliptic functions.

For several reasons it is advantageous to have available a direct method of computation, independent of auxiliary tables, based on the use of the modern high-speed calculating machine. The writer has therefore thought it desirable to set out in detail the method of calculation based on Landen's quadric transformation and the use of the scale of arithmetico-geometrical means, recapitulating briefly the original developments of Lagrange, Legendre, Gauss, Jacobi and others, extending them in several directions and keeping in mind the ultimate use of the calculating machine for numerical work.

SECTION II

HISTORICAL NOTE ON LANDEN'S TRANSFORMATION AND THE VARIOUS SCALES OF MODULI AND AMPLITUDES

The importance of realizing rapid and accurate methods of calculating the elliptic integrals now denoted by

$$u = F(\phi, k) = \int_0^\phi d\phi/\Delta(\phi, k) \text{ and } E(\phi, k) = \int_0^\phi \Delta(\phi, k) \, d\phi \dots(1),$$

where
$$\Delta(\phi, k) = \sqrt{(1 - k^2 \sin^2 \phi)},$$

was first remarked by Euler* (1766), although it was not until several years later that Landen† (1775) discovered in geometrical form the transformation which forms the basis of the existing methods of the numerical calculation of the elliptic integrals.

A method of successive transformations for the ultimate reduction of the algebraic forms of these integrals to elementary integrals was published by Lagrange‡ in 1784–85. This memoir contains an exposition of the scales of arithmetico-geometrical means, together with two types of algebraic transformations corresponding to the increasing and diminishing amplitudes of Landen's trigonometrical forms: it is shown that the given integral must lie between two limits which may be made to approach each other as closely as may be desired: the limits thus ob-

---

* Euler, *Novi Comm. Acad. Sc. Petrop.* vol. x. 1766.

† Landen, *Phil. Trans. Roy. Soc.* vol. LXV. p. 283, 1775; *Mathematical Memoirs*, London, 1780.

‡ Lagrange, "Sur une nouvelle méthode de calcul intégral pour les différentielles affectées d'un radical carré sous lequel la variable ne passe pas le quatrième degré," *Mém. de l'Acad. roy. des Sc. de Turin*, t. II. 1784–5; *Œuvres* (Gauthier-Villars, Paris, 1868), t. II. pp. 253–312.

tained for the integrals of the first and second kinds are practically identical with the limits and series obtained about the same time by Legendre.

Landen's transformation was applied by Legendre* to the numerical calculation of the elliptic integrals, and furnished the method of computation by means of which the latter's well-known tables were constructed. The scales of moduli and amplitudes employed by Legendre are briefly described below.

If we form successive moduli and amplitudes according to the recurrence formulae

$$k_{n+1} = (1 - k_n')/(1 + k_n') \quad \text{and} \quad \tan(\phi_{n+1} - \phi_n) = k_n' \tan \phi_n \quad ...(2),$$

Landen's transformation in trigonometrical form leads to the result

$$F(\phi_{n+1}, k_{n+1}) = (1 + k_n') F(\phi_n, k_n) \quad ....(3).$$

As n increases, the successive moduli k_n converge to zero, and the complementary moduli k_n' to unity, while the amplitudes increase so that $\phi_n/2^n$ tends rapidly to a definite limit: ultimately we may write $F(\phi_n, k_n) = \phi_n$ and thus obtain the result

$$F(\phi_0, k_0) = (2K/\pi) \phi_n/2^n, \quad \text{where} \quad K = \tfrac{1}{2}\pi (1 + k_1)(1 + k_2) \ldots (1 + k_n)$$
$$.........(4).$$

The result of calculating successive moduli and amplitudes in reverse order is equivalent to calculating in forward order these quantities according to the recurrence formulae

$$k_{n+1} = 2\sqrt{k_n}/(1 + k_n) \quad \text{and} \quad \sin(2\psi_{n+1} - \psi_n) = k_n \sin \psi_n \quad ...(5).$$

This transformation then gives

$$F(\psi_{n+1}, k_{n+1}) = \tfrac{1}{2}(1 + k_n) F(\psi_n, k_n) \quad ....(6).$$

As n increases the successive moduli k_{n+1} converge to unity and the amplitudes to a limit ψ_n: we thus obtain, when n is sufficiently large,

$$F(\psi_n, k_n) = \int_0^{\psi_n} \sec \psi_n d\psi_n = \log \tan(\tfrac{1}{4}\pi + \tfrac{1}{2}\psi_n) \quad ......(7),$$

giving finally

$$F(\psi_0, k_0) = \sqrt{(k_1 k_2 k_3 \ldots k_n)}/\sqrt{k_0} . \log \tan(\tfrac{1}{4}\pi + \tfrac{1}{2}\psi_n) \quad ...(8).$$

Legendre also developed for purposes of numerical calculation extremely convergent series for $E(\phi, k)$ which he employed in the tabulation of this function.

* Legendre, "Mémoire sur les intégrations par arcs d'ellipse" and "Second mémoire, etc.," *Mém. de l'Acad. des Sciences de Paris*, ann. 1786 (Paris, 1788), pp. 616–643 and 644–683; *Traité des Fonctions Elliptiques*, Paris, 1825, t. I. p. 79 *et seq.*

The method discovered independently by Gauss and called by him the "algorithm of the arithmetico-geometrical mean" originated in connection with the evaluation of complete elliptic integrals arising from a problem in attractions required in planetary theory*: in a note he mentions that the results were obtained by him as part of a more comprehensive theory, independently of the results of Lagrange and Legendre with which they are closely connected. Gauss employed a trigonometrical transformation along the following lines.

Commencing with two numbers (a_0, b_0), successive numbers (a_n, b_n) are calculated from the recurrence formulae,

$$a_{n+1} = \tfrac{1}{2}(a_n + b_n), \quad b_{n+1} = \sqrt{(a_n b_n)} \quad \ldots\ldots\ldots\ldots(9).$$

Writing $\Delta_n = \sqrt{(a_n^2 \cos^2 \phi_n + b_n^2 \sin^2 \phi_n)}$, Gauss employs the recurrence formula

$$a_n \tan \phi_{n+1} = \Delta_{n+1} \tan \phi_n,$$

or $\quad \tan \phi_{n+1} = \tan \phi_n \sqrt{(a_{n+1}/a_n)} \cdot \sqrt{(b_n + \Delta_n)}/\sqrt{(a_n + \Delta_n)} \quad \ldots(10).$

From the relation

$$\tan \phi_n = (\Delta_1 \Delta_2 \ldots \Delta_n \ldots)/(a_0 a_1 \ldots a_n \ldots) \cdot \tan \phi_0$$

we have ultimately, since the ϕ's tend to a limit ϕ_n,

$$\int_0^{\phi_0} d\phi_0/\Delta_0 = \phi_n/a_n,$$

and hence

$$a_n \int_0^{\phi_0} d\phi_0/\Delta_0 = \tan^{-1}\left[(\Delta_1 \Delta_2 \ldots \Delta_n \ldots)/(a_0 a_1 \ldots a_n \ldots) \cdot \tan \phi_0\right] \quad \ldots(11).$$

Extremely convergent series for the elliptic integral of the second kind were also obtained by Gauss. The application of this method to the numerical calculation of the elliptic integrals was considered by Jacobi some years later†.

In the course of researches carried out between 1797 and 1808, but not published until after his death considerably later‡, Gauss developed the theory of elliptic functions to a remarkable extent and derived a large number of formulae which, at a later date, were obtained independently by Abel and Jacobi and which are now associated with the

---

* Gauss, "Determinatio Attractionis quam in Punctum quodvis Positionis datae exerceret Planeta, etc.," *Comm. Gott. Soc. Reg. Scient.* iv. 1818; *Ges. Werke* (Göttingen, 1866), Bd. iii. pp. 331–355. The note mentioned above is dated Feb. 9, 1818; *Ges. Werke*, Bd. iii. pp. 357–360.

† Jacobi, *Fundamenta Nova*, 1829, sections 38 and 52; *Ges. Werke* (Berlin, 1881), Bd. i. pp. 154 and 203. Also "Numerische Berechnung der Elliptischen Functionen," *Crelle*, Bd. xxvi. pp. 93–114; *Ges. Werke* (1861), Bd. i. pp. 345–368.

‡ Gauss, *Ges. Werke* (1866), Bd. iii. pp. 361–403 and 433–469.

theory of the Theta-functions. In one of these posthumous papers, Gauss[*] employs the recurrence formula $\tan 2\phi_n = \sqrt{(b_{n+1}/a_{n+1})} \tan \phi_{n+1}$, which is much more convenient for purposes of numerical calculation than (10). The last mentioned method is considerably simpler than that of Legendre's scales of moduli, especially when a calculating machine is available, although in theory these two methods of computation are practically identical.

Although a few formulae based on the scales of arithmetico-geometrical means were briefly considered by Jacobi[†], no attempt seems to have been made by him to apply the results to actual computation.

As the method of evaluating the elliptic functions by the use of the A.G.M. scales is only briefly dealt with in existing text-books[‡], and does not seem to have been developed to the extent it deserves in various memoirs which have appeared on the subject in recent years, a tolerably full account of the subject is given in the present book, together with a brief outline of the main analytical steps. Several new formulae have been obtained by the writer which facilitate the application of this method of numerical computation and considerably extend its scope.

From the point of view of pure analysis, the modern Weierstrassian notation presents many advantages in regard to elegance and symmetry, a superiority which no longer holds, however, when it is required to proceed to numerical evaluation. As the main point of the present paper is concerned with improvements in methods of computation, the older Jacobian notation has been adhered to as far as possible, especially as it expresses most readily the elliptic functions and integrals which arise from integrations involving circular and hyperbolic functions. The results are set out in such a manner as to provide a fairly complete compendium of formulae intended to be of service to the computer in reducing to numerical results such expressions involving elliptic functions and integrals as may arise in physical and astronomical problems.

* Gauss, *loc. cit.* p. 388.

† Jacobi, *Ges. Werke* (1881), p. 357.

‡ Cayley, A., *Elliptic Functions*, 2nd Ed. (G. Bell and Sons, London, 1895), pp. 326–338; Enneper-Müller, *Elliptische Functionen* (2nd ed. 1890), pp. 361–364; Tannery-Molk, *Fonctions Elliptiques*, t. IV. 1902, pp. 269–275. According to a footnote in Halphen's *Fonctions Elliptiques*, t. II. 1888, p. 310, the subject of the arithmetico-geometrical mean was to have received special attention in the author's third volume, unfortunately incomplete at the time of his death.

SECTION III

ON THE SCALE OF ARITHMETICO-GEOMETRICAL MEANS

In forming the scale of arithmetico-geometrical means, we start with the positive numbers (a_0, b_0), of which a_0 is the greater, and form successively

$$\left. \begin{aligned} &a_1 = \tfrac{1}{2}(a_0 + b_0), & &b_1 = \sqrt{(a_0 b_0)}, & &c_1 = \tfrac{1}{2}(a_0 - b_0) \\ &a_2 = \tfrac{1}{2}(a_1 + b_1), & &b_2 = \sqrt{(a_1 b_1)}, & &c_2 = \tfrac{1}{2}(a_1 - b_1) \\ &\cdots\cdots\cdots & &\cdots\cdots\cdots & &\cdots\cdots\cdots \\ &a_n = \tfrac{1}{2}(a_{n-1} + b_{n-1}) & &b_n = \sqrt{(a_{n-1} b_{n-1})} & &c_n = \tfrac{1}{2}(a_{n-1} - b_{n-1}) \\ &\cdots\cdots\cdots\cdots, & &\cdots\cdots\cdots\cdots, & &\cdots\cdots\cdots\cdots \end{aligned} \right\} \dots(12).$$

We obviously have the relations

$$\left. \begin{aligned} a_n &= a_{n+1} + c_{n+1}, & c_n{}^2 &= a_n{}^2 - b_n{}^2 \\ b_n &= a_{n+1} - c_{n+1}, & a_n c_n &= \tfrac{1}{4} c^2_{n-1} \end{aligned} \right\} \dots\dots\dots(13).$$

The a's and b's tend to the same limit denoted by

$$M(a_0, b_0) = \lim_{n \to \infty} a_n \dots\dots\dots\dots\dots(14)$$

with extraordinary rapidity, even when a_0 and b_0 are initially numbers of very different magnitudes*.

We notice in passing that $M(a_0, b_0)$ satisfies a homogeneity relation of the form

$$\epsilon M(a_0, b_0) = M(\epsilon a_0, \epsilon b_0) \dots\dots\dots\dots\dots(15),$$

ϵ being any number. The array of numbers (12) will be referred to as the " scale of arithmetico-geometrical means (a_0, b_0)," or more briefly as the A.G.M. scale (a_0, b_0). Also, the limit $M(a_0, b_0)$ will be denoted by a_n, as long as by so doing no ambiguity is involved.

If we calculate in the same way an array of numbers commencing with $\{a_0' = a_0, b_0' = c_0 = \sqrt{(a_0{}^2 - b_0{}^2)}\}$, we derive what will be called the "complementary A.G.M. scale (a_0', b_0')," the symbols being in this case denoted by accented letters, and the limit $M(a_0', b_0')$ by a_n'.

* The usual arithmetical process of extracting square roots is easily adapted to most calculating machines of modern type. An account of these is given in *Modern Instruments and Methods of Calculation*, edited by E. M. Horseburgh (London, G. Bell and Sons, 1914); also by d'Ocagne, *Le Calcul Simplifié* (Gauthier-Villars, Paris, 1905). In carrying out the process the work is considerably shortened by keeping in mind the rule that if the first p digits out of the number n required in the square root have been obtained by the usual process, the next $p - 1$ digits can be obtained by division only, with a possible error of 1 in the last digit (Chrystal's *Algebra*, Part I. 5th ed. 1904, p. 210). To illustrate the rapid convergence of the A.G.M. scales, Gauss works out an example starting with $a_0 = 1$ and $b_0 = 0.2$: a_5 and b_5 are represented by the same number to fifteen significant figures.

It is obvious that the operations of the A.G.M. scale may be carried out backwards : if the entries are denoted by negative suffixes, they are connected with those of the complementary scale by the relations

$$a_{-n} = 2^n a_n', \qquad b_{-n} = 2^n c_n', \qquad c_{-n} = 2^n b_n' \quad \ldots\ldots\ldots(16).$$

These relations evidently hold good when accented and unaccented symbols are interchanged. As a result of these relations it will be seen that there is no advantage to be gained from considering any other than the A.G.M. scale of positive suffixes (a_n, b_n) and the corresponding complementary scale (a_n', b_n').

SECTION IV

LANDEN'S SCALE OF INCREASING AMPLITUDES

(i) *Calculation of the Elliptic Integrals of the First and Second kinds, modulus k.*

If we write $\Delta_n = \sqrt{(a_n^2 \cos^2\phi_n + b_n^2 \sin^2\phi_n)}$, the recurrence formula

$$\tan(\phi_{n+1} - \phi_n) = (b_n/a_n) \tan\phi_n \quad \ldots\ldots\ldots\ldots(17)$$

may be written in either of the forms

$$\sin(2\phi_n - \phi_{n+1}) = (c_{n+1}/a_{n+1}) \sin\phi_{n+1},$$
$$\cos(2\phi_n - \phi_{n+1}) = \Delta_{n+1}/a_{n+1} \quad \ldots\ldots(18).$$

From these we derive

$$\Delta_{n+1} + c_{n+1} \cos\phi_{n+1} = \Delta_n \quad \text{and} \quad \Delta_{n+1} - c_{n+1} \cos\phi_{n+1} = a_n b_n/\Delta_n \quad \ldots(19),$$

which give on differentiation,

$$d\phi_0/\Delta_0 = \tfrac{1}{2} d\phi_1/\Delta_1 = \tfrac{1}{4} d\phi_2/\Delta_2 = \ldots = (\tfrac{1}{2})^n d\phi_n/\Delta_n = \ldots \quad \ldots\ldots(20).$$

From the relation

$$\Delta^2_{n+1} + c_{n+1}\Delta_{n+1} \cos\phi_{n+1} = \tfrac{1}{2}\Delta_n^2 + \tfrac{1}{2} a_n b_n$$

we obtain, making use of (20),

$$\Delta_n d\phi_n - \Delta_{n+1} d\phi_{n+1} = c_{n+1} \cos\phi_{n+1} d\phi_{n+1} - a_n b_n d\phi_n/\Delta_n \quad \ldots(21).$$

Subtracting from each side the identity

$$a_n^2 d\phi_n/\Delta_n - a^2_{n+1} d\phi_{n+1}/\Delta_{n+1} = \tfrac{1}{2} c_n^2 d\phi_n/\Delta_n - a_n b_n d\phi_n/\Delta_n \ldots(22)$$

we obtain

$$(\Delta_n - a_n^2/\Delta_n) d\phi_n - (\Delta_{n+1} - a^2_{n+1}/\Delta_{n+1}) d\phi_{n+1}$$
$$= c_{n+1} \cos\phi_{n+1} d\phi_{n+1} - \tfrac{1}{2} c_n^2 d\phi_n/\Delta_n \quad \ldots(23).$$

Writing down a series of such equations, commencing with $n = 0$, and noting that as n increases, $(\Delta_n - a_n^2/\Delta_n)$ tends to zero in the limit, we have, on integrating,

$$\int_0^{\phi_0} \Delta_0 d\phi_0 - a_0{}^2 \int_0^{\phi_0} d\phi_0/\Delta_0 = c_1 \sin \phi_1 + c_2 \sin \phi_2 + \dots + c_n \sin \phi_n + \dots$$
$$- \tfrac{1}{2}(c_0{}^2 + 2c_1{}^2 + 4c_2{}^2 + \dots + 2^n c_n{}^2 + \dots) \int_0^{\phi_0} d\phi_0/\Delta_0 \dots\dots(24).$$

If we now construct the A.G.M. scale $(a_0 = 1, \ b_0 = k')$, and calculate successive amplitudes commencing with $\phi_0 = \phi$, we have $\Delta_0 = \Delta(\phi, k)$. As n increases, the a's and b's tend to the same limit a_n, while the angle $\phi_n/2^n$ tends to a finite limit. We thus obtain from (20) and (24),

$$u = F(\phi, k) = (1/a_n)(\phi_n/2^n) \dots\dots\dots\dots(25),$$
$$E(\phi, k) - F(\phi, k) = (c_1 \sin \phi_1 + c_2 \sin \phi_2 + \dots + c_n \sin \phi_n + \dots)$$
$$- \tfrac{1}{2}(c_0{}^2 + 2c_1{}^2 + 4c_2{}^2 + \dots + 2^n c_n{}^2 + \dots) F(\phi, k) \ \dots\dots(26).$$

If we commence with $\phi_0 = \tfrac{1}{2}\pi$, we have

$$\phi_1 = \pi, \ \phi_2 = 2\pi, \ \phi_3 = 4\pi \dots \phi_n/2^n = \tfrac{1}{2}\pi,$$

so that $K = \tfrac{1}{2}\pi/a_n$, and

$$(K - E)/K = \tfrac{1}{2}(c_0{}^2 + 2c_1{}^2 + 4c_2{}^2 + \dots + 2^n c_n{}^2 + \dots) \ \dots\dots(27).$$

The series (24) and (27) converge with extreme rapidity, resembling in this respect the q-series with which they are closely connected[*].

Jacobi's integral $Z(\phi, k)$ is defined by the relation

$$Z(\phi, k) = E(\phi, k) - (E/K) F(\phi, k) \dots\dots\dots(28)$$

or, if the angle ϕ is connected with the argument u by the relation

$$\sin \phi = \operatorname{sn}(u, k),$$

$$Z(u, k) = \int_0^u \operatorname{dn}^2(u, k) \, du - (E/K) u \ \dots\dots\dots(29).$$

From (26) and (27) we derive the expansion

$$Z(\phi, k) = c_1 \sin \phi_1 + c_2 \sin \phi_2 + \dots + c_n \sin \phi_n + \dots \ \ \dots(30).$$

In terms of Jacobi's Theta-function $\Theta(u, k)$,

$$\int_0^u Z(u, k) \, du = \int_0^\phi Z(\phi, k) \, d\phi/\Delta(\phi, k) = \log[\Theta(u, k)/\Theta(0, k)] \dots(31)$$

where $\qquad\qquad\qquad \Theta(0, k) = \sqrt{(2k'K/\pi)} \ \ \dots\dots\dots\dots(32).$

---

[*] These relations are discussed by Legendre (*Traité des Fonctions Elliptiques*, t. III. Deuxième Supplément, 1828, p. 111 *et seq.*). On the connection between the A.G.M. scale and the modern transformation theory of the Theta-functions, see Tannery and Molk, *Fonctions Elliptiques*, t. IV. 1902, note 2, pp. 269–273. A development of the theory of the Theta-functions based on Legendre's scales of moduli is outlined by Richelot, *Correspondenz mit Herrn Professor Schröter* (Königsberg, 1868). See Examples 19–24 of the Appendix.

From (20) we have $d\phi/\Delta(\phi, k) = (\frac{1}{2})^n d\phi_n/\Delta_n$, so that the series (30) enables us to write

$$\int_0^\phi Z(\phi, k)\, d\phi/\Delta(\phi, k) = \sum_1^\infty (c_n/2^n) \int_0^{\phi_n} \sin\phi_n\, d\phi_n/\Delta_n.$$

On integrating and making use of (19) and (18), we obtain

$$\int_0^{\phi_n} \sin\phi_n\, d\phi_n/\Delta_n = (1/c_n) \log(a_{n-1}/\Delta_{n-1}) = (1/c_n) \log\sec(2\phi_{n-2} - \phi_{n-1}),$$

leading to the expansions*

$$\int_0^u Z(u, k)\, du = \sum_1^\infty (\tfrac{1}{2})^n \log(a_{n-1}/\Delta_{n-1}) \dots\dots\dots(33),$$

or

$$\log[\Theta(u, k)/\Theta(0, k)] = \tfrac{1}{2}\log(a_0/\Delta_0) + \tfrac{1}{4}\log\sec(2\phi_0 - \phi_1)$$
$$+ \tfrac{1}{8}\log\sec(2\phi_1 - \phi_2) + \dots + (\tfrac{1}{2})^{n+1}\log\sec(2\phi_{n-1} - \phi_n) + \dots \dots(34).$$

In the series (34) it is more convenient to write

$$a_0/\Delta_0 = \cos(\phi_1 - \phi_0)/\cos\phi_0$$

for purposes of numerical calculation.

(ii) *Calculation of the Jacobian Functions* sn (u, k), cn (u, k), dn (u, k) *etc. in terms of the argument u.*

When the argument u is given and the modulus k is known, it is only necessary to compute the A.G.M. scale $(a_0 = 1, b_0 = k')$ to such a value of n that c_n is less than the small quantity determining the order of accuracy of the calculations. Since $u = F(\phi, k)$, equation (25) enables us to calculate in circular measure the angle ϕ_n from the relation

$$\phi_n = 2^n a_n u \dots\dots\dots\dots\dots\dots(35).$$

From the recurrence formula (18),

$$\sin(2\phi_{n-1} - \phi_n) = (c_n/a_n)\sin\phi_n,$$

we are enabled to calculate successively the angles $\phi_{n-1} \dots \phi_2$, ϕ_1 and finally ϕ_0, in terms of which

$$\text{sn}\,(u, k) = \sin\phi_0, \quad \text{cn}\,(u, k) = \cos\phi_0, \quad \text{dn}\,(u, k) = \Delta_0 = \cos\phi_0/\cos(\phi_1 - \phi_0)$$
$$\dots\dots(36).$$

$Z(u, k)$ may then be computed from the series (30), and $\Theta(u, k)$ from formulae (32) and (34).

---

* Formulae (30), (33) and (34) are practically the same as those given by Legendre (*Traité des Fonctions Elliptiques*, t. I. section 90), and are contained implicitly in formulae given in Gauss' posthumous papers. Formula (34) is equivalent to that given by Jacobi (*Crelle*, Bd. xxvi. pp. 93–114; *Gesammelte Werke* (1881), Bd. II. p. 357).

SECTION V

THE HYPERBOLIC SCALE OF INCREASING AMPLITUDES

(i) *Calculation of the Elliptic Integrals of the First and Second kinds, modulus k'.*

Writing $i\phi_n$ for ϕ_n throughout the formulae (17) to (24), we obtain, instead of (17),

$$\tanh(\phi_{n+1} - \phi_n) = (b_n/a_n)\tanh\phi_n \quad\dots\dots\dots\dots(37),$$

which may also be written

$$\sinh(2\phi_n - \phi_{n+1}) = (c_{n+1}/a_{n+1})\sinh\phi_{n+1}, \quad \cosh(2\phi_n - \phi_{n+1}) = \Delta_{n+1}/a_{n+1}$$
$$\dots\dots(38).$$

If we denote $\Delta_n = \sqrt{(a_n^2\cosh^2\phi_n - b_n^2\sinh^2\phi_n)}$, it is easily proved, as before, that

$$d\phi_0/\Delta_0 = \tfrac{1}{2}d\phi_1/\Delta_1 = \tfrac{1}{4}d\phi_2/\Delta_2 = \dots = (\tfrac{1}{2})^n d\phi_n/\Delta_n = \dots \quad\dots(39)$$

from which it follows that as n increases, we have in the limit,

$$\int_0^{\phi_0} d\phi_0/\Delta_0 = (1/a_n)(\phi_n/2^n).$$

If we now write

$$\sinh\phi_0 = \tan\phi \quad\dots\dots\dots\dots\dots\dots(40)$$

it is easily proved that

$$\Delta_0 = \Delta(\phi, k')/\cos\phi, \quad d\phi_0/\Delta_0 = d\phi/\Delta(\phi, k') \quad\dots\dots\dots(41),$$

so that

$$u = F(\phi, k') = (1/a_n)(\phi_n/2^n) \quad\dots\dots\dots\dots(42).$$

Formula (24) becomes,

$$\int_0^{\phi_0}\Delta_0 d\phi_0 - a_0^2\int_0^{\phi_0}d\phi_0/\Delta_0 = (c_1\sinh\phi_1 + c_2\sinh\phi_2 + \dots + c_n\sinh\phi_n + \dots)$$
$$-\tfrac{1}{2}(c_0^2 + 2c_1^2 + 4c_2^2 + \dots + 2^n c_n^2 + \dots)\int_0^{\phi_0}d\phi_0/\Delta_0 \quad\dots(43).$$

It is easily shown that

$$\int_0^{\phi_0}\Delta_0 d\phi_0 = \int_0^{\phi}\Delta(\phi, k')d\phi/\cos^2\phi$$
$$= \int_0^u \mathrm{dn}^2(u, k')\,du/\mathrm{cn}^2(u, k') = \int_0^u \mathrm{dn}^2(iu, k)\,du,$$

or in terms of the Z-function of imaginary argument,

$$\int_0^{\phi_0}\Delta_0 d\phi_0 = (1/i)Z(iu, k) + (E/K)u.$$

Substituting in (43), and making use of (27), we derive the series, corresponding to that of (30),

$$(1/i)Z(iu, k) = c_1\sinh\phi_1 + c_2\sinh\phi_2 + \dots + c_n\sinh\phi_n + \dots \quad\dots(44).$$

The following formula, corresponding to (34), easily follows by integration,

$$\log\left[\Theta\left(iu, k\right)/\Theta\left(0, k\right)\right] = \tfrac{1}{2}\log\left(a_0/\Delta_0\right) + \sum_{1}^{\infty}\left(\tfrac{1}{2}\right)^{n+1}\log\operatorname{sech}\left(2\phi_{n-1} - \phi_n\right)$$

$$\ldots\ldots(45)$$

in which we may conveniently write $a_0/\Delta_0 = \cosh\left(\phi_1 - \phi_0\right)/\cosh\phi_0$. From (42) we easily deduce the relations,

$$\left.\begin{aligned}(1/i)\operatorname{sn}\left(iu, k\right) &= \operatorname{sn}\left(u, k'\right)/\operatorname{cn}\left(u, k'\right) = \sinh\phi_0 \\ \operatorname{cn}\left(iu, k\right) &= \qquad 1/\operatorname{cn}\left(u, k'\right) = \cosh\phi_0 \\ \operatorname{dn}\left(iu, k\right) &= \operatorname{dn}\left(u, k'\right)/\operatorname{cn}\left(u, k'\right) = \Delta_0 = \cosh\phi_0/\cosh\left(\phi_1 - \phi_0\right)\end{aligned}\right\}$$

$$\ldots\ldots(46).$$

Making use of Jacobi's well-known relation in the form

$$Z\left(u, k'\right) = \operatorname{sn}\left(u, k'\right)\operatorname{dn}\left(u, k'\right)/\operatorname{cn}\left(u, k'\right) - \pi u/\left(2KK'\right) + iZ\left(iu, k\right)$$

$$\ldots\ldots(47)$$

we have from (28) and (46)

$$\mathfrak{Z}\left(\phi, k'\right) - \left(E'/K'\right)F\left(\phi, k'\right)$$
$$= \sinh\phi_0/\cosh\left(\phi_1 - \phi_0\right) - \pi F\left(\phi, k'\right)/\left(2KK'\right) + iZ\left(iu, k\right).$$

Noting that Legendre's relation gives

$$E/K + E'/K' - 1 = \tfrac{1}{2}\pi/\left(KK'\right)\quad\ldots\ldots\ldots(48)$$

we obtain from (27)

$$E\left(\phi, k'\right) = \sinh\phi_0/\cosh\left(\phi_1 - \phi_0\right) + \left(1 - E/K\right)F\left(\phi, k'\right) + iZ\left(iu, k\right),$$

or, finally, making use of (27), (42) and (44),

$$E\left(\phi, k'\right) = \sinh\phi_0/\cosh\left(\phi_1 - \phi_0\right)$$
$$+ \tfrac{1}{2}\left(c_0^2 + 2c_1^2 + \ldots + 2^n c_n^2 + \ldots\right)\left(1/a_n\right)\left(\phi_n/2^n\right)$$
$$- \left(c_1\sinh\phi_1 + c_2\sinh\phi_2 + \ldots + c_n\sinh\phi_n + \ldots\right)\quad\ldots(49).$$

(ii) *Calculation of the Jacobian Functions* $\operatorname{sn}\left(u, k'\right)$, $\operatorname{cn}\left(u, k'\right)$, $\operatorname{dn}\left(u, k'\right)$, *etc. in terms of the argument* u.

Compute the A.G.M. scale ($a_0 = 1$, $b_0 = k'$) and calculate ϕ_n from the relation $\phi_n = 2^n a_n u$, n being such that c_n is less than the small quantity determining the order of accuracy of the calculations. The hyperbolic recurrence formula

$$\sinh\left(2\phi_{n-1} - \phi_n\right) = \left(c_n/a_n\right)\sinh\phi_n$$

enables us to calculate successively $\phi_{n-1} \ldots \phi_2, \phi_1$, and finally ϕ_0, in terms of which we have from (46)

$$\operatorname{sn}\left(u, k'\right) = \tanh\phi_0, \quad \operatorname{cn}\left(u, k'\right) = 1/\cosh\phi_0, \quad \operatorname{dn}\left(u, k'\right) = 1/\cosh\left(\phi_1 - \phi_0\right)$$

$$\ldots\ldots(50).$$

$Z\left(u, k'\right)$ may then be calculated from (47) and (44).

(iii) *Calculation of the Jacobian Functions of imaginary argument,* sn (iu, k), cn (iu, k), dn (iu, k), etc.

The procedure outlined above, together with equations (46), obviously enables us to calculate the Jacobian functions of imaginary argument and modulus k, while formulae (44) and (45) serve for the functions $Z(iu, k)$ and $\Theta(iu, k)$.

SECTION VI

GAUSS' SCALE OF INCREASING AMPLITUDES

(i) *Circular Scale.*

In connection with the A.G.M. scale, Gauss made use of the recurrence formula *

$$\tan \chi_{n+1} = \sqrt{(a_{n+1}/b_{n+1})} \tan 2\chi_n \quad \ldots\ldots\ldots\ldots(51).$$

It is easily proved that the above formula is connected with Landen's scale (17) by the relation

$$\tan \chi_n = \sqrt{(b_n/a_n)} \tan \phi_n \ldots\ldots\ldots\ldots\ldots\ldots(52)$$

or

$$\phi_{n+1} - \phi_n = 2\chi_{n-1} \ldots\ldots\ldots\ldots\ldots\ldots\ldots(53).$$

In the limit, as n tends to infinity, (b_n/a_n) approaches unity, so that χ_n and ϕ_n both tend to the same limit. It thus follows from (25) that if we form the A.G.M. scale $(a_0 = 1, b_0 = k')$, and starting with χ_0 determined by $\tan \chi_0 = \sqrt{(b_0/a_0)} \tan \phi$, compute successively $\chi_1, \chi_2, \cdots \chi_n \cdots$, we have

$$F(\phi, k) = (1/a_n)(\chi_n/2^n) \quad \ldots\ldots\ldots\ldots(54).$$

From (52) and (53) it is a simple matter to derive successively $\phi_1, \phi_2 \ldots \phi_n \ldots$ for use in the formulae of Section IV. When these angles are required, it is, however, simpler to make direct use of Landen's recurrence formula (17) or (18).

(ii) *Hyperbolic Scale.*

It is evident that formulae (51) and (52) may be written in the hyperbolic form

$$\tanh \chi_{n+1} = \sqrt{(a_{n+1}/b_{n+1})} \tanh 2\chi_n \ldots\ldots\ldots\ldots(55),$$

$$\tanh \chi_n = \sqrt{(b_n/a_n)} \tanh \phi_n \quad \ldots\ldots\ldots\ldots(56).$$

* Gauss, *Ges. Werke* (1866), Bd. III. p. 388. To compare Gauss' results with those of the present paper we note that $\phi_n = 2^n V^n$, $\chi_n = 2^n U^{n+1}$. It then appears (p. 392) that formula (26) of the present book was also obtained by Gauss, independently of Legendre (foot-note on p. 9).

As n increases indefinitely, χ_n and ϕ_n tend to the same limit, so that if we write

$$\tan \phi = \sinh \phi_0 \quad \text{or} \quad \sin \phi = \tanh \phi_0 \quad\ldots\ldots\ldots\ldots(57)$$

it follows from (42) that

$$F(\phi, k') = (1/a_n)(\phi_n/2^n) = (1/a_n)(\chi_n/2^n)\ldots\ldots\ldots\ldots(58).$$

Noting that, as before,

$$\phi_{n+1} - \phi_n = 2\chi_{n-1} \quad\ldots\ldots\ldots\ldots\ldots\ldots(59)$$

it is a simple matter to derive successively ϕ_1, ϕ_2, $\ldots$ ϕ_n $\ldots$ for use in the formulae of Section V. In such cases, however, it is simpler to make direct use of the hyperbolic recurrence formulae (37) or (38).

(iii) *Calculation of K', E' and the nome $q = e^{-\pi K'/K}$.*

If we write $\phi = \frac{1}{2}\pi$, it follows from (57) that $\tanh \phi_0 = 1$, and from (56) that $\tanh \chi_0 = \sqrt{(b_0/a_0)}$. From (55) we deduce that

$$\tanh 2\chi_0 = b_1/a_1, \quad \tanh 2\chi_1 = b_2/a_2, \quad \ldots \tanh 2\chi_n = b_{n+1}/a_{n+1},$$

and thus $\quad 2\chi_n = \frac{1}{2}\log\{(a_{n+1}+b_{n+1})/(a_{n+1}-b_{n+1})\} = \frac{1}{2}\log(a_{n+2}/c_{n+2})$

or $\quad\quad\quad (\chi_n/2^n) = (\frac{1}{2})^{n+2}\log(a_{n+2}/c_{n+2}).$

It follows from (58) that as n is indefinitely increased, $(\frac{1}{2})^n \log(a_n/c_n)$ tends to a finite limit given by

$$a_n F(\tfrac{1}{2}\pi, k') = \tfrac{1}{2}\pi K'/K = \tfrac{1}{2}\log(1/q) = (\tfrac{1}{2})^n \log(a_n/c_n) \ldots(60).$$

Making use of the relation (13), $a_n c_n = \frac{1}{4}c^2_{n-1}$, we have

$$\log(a_n/c_n) = 2\log(2a_n/a_{n-1}) + 2\log(a_{n-1}/c_{n-1}),$$

which gives by successive application,

$$\log(a_n/c_n) =$$
$$2\log(2a_n/a_{n-1}) + 4\log(2a_{n-1}/a_{n-2}) + \ldots + 2^n\log(2a_1/a_0) + 2^n\log(a_0/c_0).$$

Inverting the order of the series, we obtain

$$(\tfrac{1}{2})^n \log(a_n/c_n) = \log(a_0/c_0) + \log 2\{1 + \tfrac{1}{2} + \tfrac{1}{4} + \ldots + (\tfrac{1}{2})^{n-1}\}$$
$$+ \log(a_1/a_0) + \tfrac{1}{2}\log(a_2/a_1) + \ldots + (\tfrac{1}{2})^{n-1}\log(a_n/a_{n-1}).$$

As n is increased indefinitely we obtain the series

$$a_n K' = \tfrac{1}{2}\log(4a_1/c_1) - \sum_{1}^{\infty}(\tfrac{1}{2})^n \log(a_n/a_{n+1}) \quad\ldots\ldots\ldots(61).$$

Making use of Legendre's relation (48) in the form

$$E' = (1 - E/K)K' + a_n$$

and using (27), we obtain the following useful formulae,

$$E' = \tfrac{1}{4}(1/a_n)(c_0{}^2 + 2c_1{}^2 + 4c_2{}^2 + \ldots + 2^n c_n{}^2 + \ldots)$$
$$\times \{\log(4a_1/c_1) - \log(a_1/a_2) - \tfrac{1}{2}\log(a_2/a_3) - \ldots\} + a_n \quad\ldots\ldots(62),$$

$$K' - E' = \tfrac{1}{2}(1/a_n)(1 - \tfrac{1}{2}c_0{}^2 - c_1{}^2 - 2c_2{}^2 - \ldots - 2^{n-1}c_n{}^2 - \ldots)$$
$$\times \{\log(4a_1/c_1) - \log(a_1/a_2) - \tfrac{1}{2}\log(a_2/a_3) - \ldots\} - a_n \quad\ldots\ldots(63)$$

for the calculation of the complete elliptic integrals of complementary modulus in terms of the A.G.M. scale ($a_0 = 1$, $b_0 = k'$).

The nome q is defined by the relation $\log(1/q) = \pi K'/K$, from which it follows that

$$\log q = \sum_1^\infty (\tfrac{1}{2})^{n-1} \log(a_n/a_{n+1}) - \log(4a_1/c_1) \quad \ldots\ldots(64)$$

which is more convenient for numerical computation than (60). It may be noted that the entire series of calculations may be carried out with ordinary logarithms to the base 10, enabling q and powers of q to be evaluated directly*.

The complementary *nome* $q' = e^{-\pi K/K'}$ is obviously related to q by the formula

$$\log q \cdot \log q' = \pi^2.$$

SECTION VII

LANDEN'S SCALE OF DECREASING AMPLITUDES

(i) *Calculation of the Elliptic Integrals of the First and Second kinds, modulus k'.*

If we denote

$$\Delta_n = \sqrt{(a_n{}^2 \cos^2 \psi_n + c_n{}^2 \sin^2 \psi_n)} \quad \ldots\ldots\ldots(65),$$

the recurrence formula

$$\tan(\psi_n - \psi_{n+1}) = (c_{n+1}/a_{n+1}) \tan \psi_{n+1} \quad \ldots\ldots\ldots(66)$$

may be written in either of the forms

$$\sin(2\psi_{n+1} - \psi_n) = (b_n/a_n) \sin \psi_n, \quad \cos(2\psi_{n+1} - \psi_n) = \Delta_n/a_n \quad \ldots(67).$$

From these we deduce

$$\Delta_n + b_n \cos \psi_n = 2\Delta_{n+1}, \quad \Delta_n - b_n \cos \psi_n = 2a_{n+1}c_{n+1}/\Delta_{n+1} \quad \ldots(68),$$

which give by differentiation

$$d\psi_0/\Delta_0 = d\psi_1/\Delta_1 = d\psi_2/\Delta_2 = \ldots = d\psi_n/\Delta_n = \ldots \quad \ldots\ldots(69).$$

If we construct the A.G.M. scale ($a_0 = 1$, $b_0 = k'$) we notice that as n increases, ψ_n tends rapidly to a finite limit. If we start with $\psi_0 = \psi$, $\Delta_0 = \Delta(\psi, k')$, we have, since c_n ultimately tends to zero,

---

* The limit derived in (60) does not appear to be generally known, although an equivalent formula is given by Legendre (*Fonctions Elliptiques*, t. I. pp. 94, 101). Although formula (60) does not appear to have been explicitly stated by Gauss, the series (61) is found among a number of formulae given in his posthumous papers (Gauss, *Ges. Werke*, Bd. III. p. 377). Neither (60) nor (61) appears to have been noticed by Jacobi or by subsequent writers on elliptic functions.

$$u = F(\psi, k') = (1/a_n) \int_0^{\psi_n} d\psi_n / \cos \psi_n = (1/a_n) \log \tan (\tfrac{1}{4}\pi + \tfrac{1}{2}\psi_n) \;...(70),$$

a useful formula due to Legendre*.

From (68) we easily obtain the relation

$$\Delta_n^2 + b_n \Delta_n \cos \psi_n = 2\Delta^2_{n+1} + \tfrac{1}{2}c_n^2$$

which gives

$$\Delta_n d\psi_n - 2\Delta_{n+1} d\psi_{n+1} = -b_n \cos \psi_n d\psi_n + \tfrac{1}{2}c_n^2 d\psi_n / \Delta_n.$$

Writing down the system of equations for $n = 0, 1, 2, \ldots n$, multiplying successively by $1, 2, 4, \ldots 2^n$, adding and integrating, we obtain the result

$$\int_0^{\psi_0} \Delta_0 d\psi_0 - 2^{n+1} \int_0^{\psi_{n+1}} \Delta_{n+1} d\psi_{n+1} = \tfrac{1}{2}(c_0^2 + 2c_1^2 + 4c_2^2 + \ldots + 2^n c_n^2) \int_0^{\psi_0} d\psi_0 / \Delta_0$$
$$- (b_0 \sin \psi_0 + 2b_1 \sin \psi_1 + \ldots + 2^n b_n \sin \psi_n) \;\ldots\ldots(71).$$

Adding to each side the identity

$$2^{n+1} \int_0^{\psi_{n+1}} b_{n+1} \cos \psi_{n+1} d\psi_{n+1} = b_{n+1} \sin \psi_{n+1} + (2^{n+1} - 1) b_{n+1} \sin \psi_{n+1}$$

and re-arranging the terms, (71) may be written

$$\int_0^{\psi_0} \Delta_0 d\psi_0 - 2^{n+1} \int_0^{\psi_{n+1}} (\Delta_{n+1} - b_{n+1} \cos \psi_{n+1}) \, d\psi_{n+1}$$
$$= \tfrac{1}{2}(c_0^2 + 2c_1^2 + \ldots + 2^n c_n^2) \int_0^{\psi_0} d\psi_0 / \Delta_0 + b_{n+1} \sin \psi_{n+1} + (b_1 \sin \psi_1 - b_0 \sin \psi_0)$$
$$+ 3(b_2 \sin \psi_2 - b_1 \sin \psi_1) + \ldots + (2^{n+1} - 1)(b_{n+1} \sin \psi_{n+1} - b_n \sin \psi_n)$$
$$\ldots\ldots(72).$$

It is not difficult to establish the identity

$$b_{n+1} \sin \psi_{n+1} - b_n \sin \psi_n = 2c_{n+2} \tan (2\psi_{n+2} - \psi_{n+1}) / \cos (2\psi_{n+3} - \psi_{n+2})$$
$$\ldots\ldots(73).$$

As n increases indefinitely, a_{n+1} and b_{n+1} tend to the same limit a_n, while c_{n+1} tends to zero. It follows that the second integral on the left-hand side of (72) tends to zero, so that we finally obtain, making use of (70),

$$E(\psi, k') = a_n \sin \psi_n + \tfrac{1}{2}(c_0^2 + 2c_1^2 + \ldots + 2^n c_n^2 + \ldots)$$
$$\times (1/a_n) \log \tan (\tfrac{1}{4}\pi + \tfrac{1}{2}\psi_n) + \underset{2}{\Sigma} (\psi_n, c_n) \;\ldots\ldots(74),$$

where for the sake of brevity we denote

$$\underset{2}{\Sigma} (\psi_n, c_n) = 2c_2 \frac{\tan (2\psi_2 - \psi_1)}{\cos (2\psi_3 - \psi_2)} + 6c_3 \frac{\tan (2\psi_3 - \psi_2)}{\cos (2\psi_4 - \psi_3)} + \ldots$$
$$+ 2(2^n - 1) c_{n+2} \frac{\tan (2\psi_{n+2} - \psi_{n+1})}{\cos (2\psi_{n+3} - \psi_{n+2})} + \ldots \;\ldots\ldots(75),$$

* Legendre, *Fonctions Elliptiques*, t. I. chap. XIX.

a useful formula for the numerical calculation of the second elliptic integral*.

Making use of (27), (28), (48) and (70), we obtain from (74) the following interesting expressions

$$Z(\psi, k') + \tfrac{1}{2}\pi\, F(\psi, k')/(KK') = a_n \sin\psi_n + \sum_2 (\psi_n, c_n) \quad \ldots(76),$$

or, in terms of u, by (70) we have

$$\sin\psi_n = \tanh(a_n u),$$

and
$$Z(u, k') + a_n u/K' = a_n \tanh(a_n u) + \sum_2 (\psi_n, c_n) \Bigg\} \quad \ldots(77).$$

Jacobi's relation (47), taken with equation (76), gives

$$(1/i)\, Z(iu, k) = \tan\psi_0 \cos(2\psi_1 - \psi_0) - a_n \sin\psi_n - \sum_2 (\psi_n, c_n) \quad (78).$$

The addition theorem

$$Z(u) + Z(v) - Z(u+v) = k^2 \operatorname{sn}(u+v)\operatorname{sn} u \operatorname{sn} v \quad \ldots\ldots(79),$$

with iu for u and $v = \pm(K+iK')$ gives two interesting formulae

$$iZ(iu + K + iK', k) = a_n(1 + \sin\psi_n) + \sum_2 (\psi_n, c_n) \Bigg\}$$
$$(1/i)\, Z(iu - K - iK', k) = a_n(1 - \sin\psi_n) - \sum_2 (\psi_n, c_n) \Bigg\} \quad \ldots(80).$$

If we notice that

$$\int_0^{\psi_n} \sin\psi_n\, d\psi_n/\Delta_n = (1/b_n)\log(a_{n+1}/\Delta_{n+1}) = (1/b_n)\log\sec(2\psi_{n+2} - \psi_{n+1}),$$

the result of integrating (72) written in the form

$$Z(u, k') + a_n u/K' = a_n \sin\psi_n + \sum_0^\infty (2^{n+1} - 1)(b_{n+1}\sin\psi_{n+1} - b_n\sin\psi_n)$$

gives rise to a formula for the numerical calculation of $\Theta(u, k')$,

$$\log\left[\frac{\Theta(u, k')}{\Theta(0, k')}\right] + \tfrac{1}{2}a_n u^2/K'$$
$$= \log\sec\psi_n + \sum_0^\infty (2^{n+1} - 1)\log\left[\frac{\cos(2\psi_{n+2} - \psi_{n+1})}{\cos(2\psi_{n+3} - \psi_{n+2})}\right] \quad \ldots\ldots(81),$$

in which, if desirable, we may write $\sec\psi_n = \cosh(a_n u)$.

(ii) *Calculation of the Jacobian Functions* $\operatorname{sn}(u, k')$, $\operatorname{cn}(u, k')$, $\operatorname{dn}(u, k')$, *etc. in terms of the argument* u.

Form the A.G.M. scale $(a_0 = 1,\ b_0 = k')$ and determine ψ_n from $\sin\psi_n = \tanh(a_n u)$, n being such that c_n is less than the small quantity determining the order of accuracy of the calculations.

* The series (71) was obtained by Legendre (*Fonctions Elliptiques*, t. I. chap. XXI.) and left by him in this non-convergent form. As far as the writer is aware, the convergent series (75) is new.

From the recurrence formula (66), $\tan(\psi_{n-1} - \psi_n) = (c_n/a_n)\tan\psi_n$, compute $\psi_{n-1} \ldots \psi_2, \psi_1$, and finally ψ_0, in terms of which

$$\text{sn}(u, k') = \sin\psi_0, \quad \text{cn}(u, k') = \cos\psi_0, \quad \text{dn}(u, k') = \Delta_0 = \cos(2\psi_1 - \psi_0)$$
$$\ldots\ldots(82);$$

$Z(u, k')$ may then be calculated from (77) and $\Theta(u, k')$ from (81).

SECTION VIII

THE HYPERBOLIC SCALE OF DECREASING AMPLITUDES

(i) *Calculation of the Elliptic Integrals of the First and Second kinds, modulus k.*

If we write $i\psi$ for ψ throughout the formulae (65) to (69), (65) becomes

$$\Delta_n = \sqrt{(a_n^2 \cosh^2\psi_n - c_n^2 \sinh^2\psi_n)}\ldots\ldots\ldots(83),$$

while the recurrence formula (66) may be written

$$\tanh(\psi_n - \psi_{n+1}) = (c_{n+1}/a_{n+1})\tanh\psi_{n+1}\ \ldots\ldots(84),$$

or, more conveniently,

$$\sinh(2\psi_{n+1} - \psi_n) = (b_n/a_n)\sinh\psi_n, \quad \cosh(2\psi_{n+1} - \psi_n) = \Delta_n/a_n\ \ldots(85).$$

As in Section VII (i), it is readily proved that

$$d\psi_0/\Delta_0 = d\psi_1/\Delta_1 = d\psi_2/\Delta_2 = \ldots = d\psi_n/\Delta_n = \ldots\ \ldots\ldots(86).$$

If we construct the A.G.M. scale $(a_0 = 1,\ b_0 = k')$ and write

$$\tan\psi = \sinh\psi_0\ \ldots\ldots\ldots\ldots\ldots(87),$$

is is readily proved that

$$\Delta_0 = \Delta(\psi, k)/\cos\psi \quad \text{and} \quad d\psi_0/\Delta_0 = d\psi/\Delta(\psi, k)\ \ldots\ldots(88).$$

Thus, writing $u = F(\psi, k)$, we have

$$\left.\begin{array}{l}\text{sn}(u, k) = \sin\psi = \tanh\psi_0, \quad \text{cn}(u, k) = \cos\psi = \text{sech}\,\psi_0 \\ \text{dn}(u, k) = \Delta(\psi, k) = \Delta_0/\cosh\psi_0 = \cosh(2\psi_1 - \psi_0)/\cosh\psi_0\end{array}\right\}\ \ldots(89).$$

As n increases, the ψ's tend to a limit ψ_n, while c_n tends to zero. Thus the result of integrating (86) gives

$$u = F(\psi, k) = (1/a_n)\int_0^{\psi_n} d\psi_n/\cosh\psi_n = (1/a_n)\tan^{-1}(\sinh\psi_n)\ \ldots(90).$$

If we notice that

$$\int_0^{\psi_0}\Delta_0\,d\psi_0 = \int_0^{\psi}\Delta(\psi, k)\,d\psi/\cos^2\psi$$
$$= \int_0^u \text{dn}^2(u, k)\,du/\text{cn}^2(u, k) = \int_0^u \text{dn}^2(iu, k')\,du,$$

we have from (29)

$$\int_0^{\psi_0} \Delta_0 \, d\psi_0 = (1/i) \, Z \, (iu, \, k') + (E'/K') \, u.$$

It is easily seen that equation (72) holds good when $i\psi_n$ is written for ψ_n. Hence if we write

$$\sideset{}{'}\sum_2 (\psi_n, c_n) = 2c_2 \frac{\tanh (2\psi_2 - \psi_1)}{\cosh (2\psi_3 - \psi_2)} + 6c_3 \frac{\tanh (2\psi_3 - \psi_2)}{\cosh (2\psi_4 - \psi_3)} + \cdots$$
$$+ 2 \, (2^n - 1) \, c_{n+2} \frac{\tanh (2\psi_{n+2} - \psi_{n+1})}{\cosh (2\psi_{n+3} - \psi_{n+2})} + \cdots \quad \cdots (91),$$

and make use of (47) and (48), we obtain after some reductions

$$Z \, (\psi, \, k) = \tanh \psi_0 \cosh (2\psi_1 - \psi_0) - a_n \sinh \psi_n - \sideset{}{'}\sum_2 (\psi_n, c_n) \quad \cdots (92).$$

The relations (27) and (28) enable us to calculate the second elliptic integral $E \, (\psi, \, k)$.

(ii) *Calculation of the Jacobian Functions* sn $(u, \, k)$, cn $(u, \, k)$, dn $(u, \, k)$, *etc. in terms of the argument u.*

Compute the A.G.M. scale $a_0 = 1$, $b_0 = k'$ and determine ψ_n from (90), which may be written

$$\sinh \psi_n = \tan (a_n u) \quad \cdots\cdots\cdots\cdots\cdots (93),$$

n being such that c_n is less than the small quantity determining the order of accuracy of the calculations.

We then calculate the amplitudes $\psi_{n+1} \cdots \psi_2, \psi_1, \psi_0$ from the recurrence formula

$$\tanh (\psi_{n-1} - \psi_n) = (c_n/a_n) \tanh \psi_n.$$

Equation (89) then gives

$$\left. \begin{array}{l} \text{sn} \, (u, \, k) = \tanh \psi_0, \quad \text{cn} \, (u, \, k) = \text{sech} \, \psi_0 \\ \text{dn} \, (u, \, k) = \cosh (2\psi_1 - \psi_0)/\cosh \psi_0 \end{array} \right\} \quad \cdots\cdots\cdots (94).$$

Equations (91) and (92) then enable us to calculate $Z \, (u, \, k)$. To compute $\Theta \, (u, \, k)$, we note that (92) may also be written

$$Z \, (u, \, k) = \text{sn} \, (u, \, k) \, \text{dn} \, (u, \, k)/\text{cn} \, (u, \, k) - a_n \sinh \psi_n$$
$$- \sum_0^\infty (2^{n+1} - 1) \, (b_{n+1} \sinh \psi_{n+1} - b_n \sinh \psi_n).$$

Integrating this expression with respect to u and noticing that

$$\int_0^{\psi_n} \sinh \psi_n \, d\psi_n / \Delta_n = (1/b_n) \log \cosh (2\psi_{n+2} - \psi_{n+1}),$$

we obtain the following series for $\Theta \, (u, \, k)$,

$$\log \left[\frac{\Theta \, (u, \, k)}{\Theta \, (0, \, k)} \right] = \log \left[\frac{\cosh \psi_0}{\cosh \psi_n} \right] + \sum_0^\infty (2^{n+1} - 1) \log \left[\frac{\cosh (2\psi_{n+2} - \psi_{n+1})}{\cosh (2\psi_{n+3} - \psi_{n+2})} \right]$$
$$\cdots\cdots (95).$$

SECTION IX

ON THE NUMERICAL COMPUTATION OF THE THIRD ELLIPTIC INTEGRAL

As defined by Legendre, the third elliptic integral may be written

$$\Pi_3 (n, k, \phi) = \int_0^\phi \frac{d\phi}{(1 + n \sin^2 \phi) \Delta (\phi, k)} \cdots\cdots\cdots(96).$$

The parameter n is taken as real, and may take any value, positive or negative.

Writing $u = F(\phi, k)$, $n = -k^2 \operatorname{sn}^2(a, k)$, Jacobi considers the integral

$$\Pi (u, a) = \int_0^u \frac{k^2 \operatorname{sn} a \operatorname{cn} a \operatorname{dn} a \operatorname{sn}^2 u \, du}{1 - k^2 \operatorname{sn}^2 a \operatorname{sn}^2 u} = u Z(a) + \tfrac{1}{2} \log \left[\frac{\Theta(a - u)}{\Theta(a + u)} \right] \dots(97).$$

Legendre's integral is connected with Jacobi's by the relation

$$\Pi_3 (n, k, \phi) = u + \operatorname{sn} a / (\operatorname{cn} a \operatorname{dn} a) . \Pi (u, a) \quad \dots\dots(98),$$

and in this book is differentiated from it by the suffix $_3$.

In order to include all real values of n in (96), a is not restricted to be real, and the treatment is usually divided into four cases:

Case I. n negative between 0 and $-k^2$ } *Hyperbolic Cases.*
Case II. n negative between -1 and $-\infty$ }

Case III. n negative between $-k^2$ and -1} *Circular Cases.*
Case IV. n positive between 0 and $+\infty$ }

CASE I. *n negative, between 0 and* $-k^2$. *Hyperbolic case.*

We write $n = -k^2 \operatorname{sn}^2(a, k) = -k^2 \sin^2 \theta$, where $0 < a < K$ ...(99).

From (97) and (98) we have

$$\Pi_3 (n, k, \phi) = u \left[1 + \frac{\operatorname{sn}(a, k) Z(a, k)}{\operatorname{cn}(a, k) \operatorname{dn}(a, k)} \right]$$
$$+ \tfrac{1}{2} \frac{\operatorname{sn}(a, k)}{\operatorname{cn}(a, k) \operatorname{dn}(a, k)} \log \left[\frac{\Theta(a - u, k)}{\Theta(a + u, k)} \right] \dots\dots(100).$$

Compute the A.G.M. scale ($a_0 = 1$, $b_0 = k'$), and starting with $\phi_0 = \theta$, make use of the recurrence formula (17), $\tan(\phi_{n+1} - \phi_n) = (b_n/a_n) \tan \phi_n$. We then have $\operatorname{sn}(a, k) = \sin \phi_0$, $\operatorname{cn}(a, k) = \cos \phi_0$, $\operatorname{dn}(a, k) = \cos \phi_0 / \cos(\phi_1 - \phi_0)$, and $Z(a, k) = c_1 \sin \phi_1 + c_2 \sin \phi_2 + \dots + c_n \sin \phi_n + \dots$, while $a_n a = \phi_n / 2^n$. We thus have

$$\Pi_3 (n, k, \phi)$$
$$= u \left[1 + \frac{\tan \phi_0 \cos(\phi_1 - \phi_0)}{\cos \phi_0} \{ c_1 \sin \phi_1 + c_2 \sin \phi_2 + \dots + c_n \sin \phi_n \dots \} \right]$$
$$+ \tfrac{1}{2} \frac{\tan \phi_0 \cos(\phi_1 - \phi_0)}{\cos \phi_0} \log \left[\frac{\Theta(a - u, k)}{\Theta(a + u, k)} \right] \dots\dots(101).$$

The calculation of u by the recurrence formula (17), commencing with $\phi_0 = \phi$, presents no difficulty. The most convenient method of computing the last term of (101) is to make use of Jacobi's series

$$\Theta(x) = 1 - 2q \cos 2a_n x + 2q^4 \cos 4a_n x - 2q^9 \cos 6a_n x + \ldots \quad \ldots(102)$$

in which we have written $a_n = \frac{1}{2}\pi/K$. The nome q is most conveniently calculated from (64), $\log q = \sum_1^\infty (\frac{1}{2})^{n-1} \log (a_n/a_{n+1}) - \log (4a_1/c_1)$.

If we write $\phi = \frac{1}{2}\pi$, the last term of (101) vanishes, and we obtain for the complete elliptic integral of the third kind the expression

$$\Pi_3(n, k, \tfrac{1}{2}\pi) = K[1 + \tan\phi_0 \sec\phi_0 \cos(\phi_1 - \phi_0)\{c_1 \sin\phi_1 + c_2 \sin\phi_2 + \ldots + c_n \sin\phi_n + \ldots\}] \quad \ldots\ldots(103).$$

We may also make use of the hyperbolic scale of decreasing amplitudes and the formulae of § 8. The final results are not, however, so convenient for computation as (101) and (103).

CASE II. *n negative, between -1 and $-\infty$. Hyperbolic case.*

We write

$$n = -k^2 \operatorname{sn}^2(a + iK', k) = -1/\operatorname{sn}^2(a, k) = -1/\sin^2\theta, \text{ where } 0 < a < K \quad \ldots\ldots(104).$$

Equation (98) becomes

$$\Pi_3(n, k, \phi) = \operatorname{sn}^2 a \int_0^u \frac{du}{\operatorname{sn}^2 a - \operatorname{sn}^2 u} = u - \frac{\operatorname{sn} a}{\operatorname{cn} a \operatorname{dn} a} \Pi(u, a + iK') \quad \ldots(105).$$

We have

$$Z(a + iK') = Z(a) + \operatorname{cn} a \operatorname{dn} a/\operatorname{sn} a - \tfrac{1}{2}\pi i/K \quad \ldots\ldots(106)$$

and

$$\Theta(z + iK') = iBH(z) \quad \ldots\ldots\ldots\ldots\ldots\ldots\ldots(107),$$

where

$$\log B = \tfrac{1}{4}\pi K'/K - \tfrac{1}{2}\pi i z/K.$$

$H(z)$ is Jacobi's Theta-function given by the expansion *

$$H(z) = 2q^{\frac{1}{4}}(\sin a_n z - q^2 \sin 3a_n z + q^6 \sin 5a_n z - q^{12} \sin 7a_n z + \ldots) \quad \ldots\ldots(108).$$

Writing successively $(a - u)$ and $(a + u)$ for z in (107), and $a + iK$ for a in (97), we obtain

$$\Pi(u, a + iK') = uZ(a, k) + u \operatorname{cn} a \operatorname{dn} a/\operatorname{sn} a + \tfrac{1}{2}\log[H(a - u)/H(a + u)],$$

and hence from (105), $u < a$,

$$\Pi_3(n, k, \phi) = -\operatorname{sn} a/(\operatorname{cn} a \operatorname{dn} a)[uZ(a, k) + \tfrac{1}{2}\log\{H(a - u)/H(a + u)\}] \quad \ldots\ldots(109).$$

* Jacobi, *Fundamenta Nova*, §§ 62, 63; *Ges. Werke*, 1881, Bd. II. pp. 229–231; Cayley, *Elliptic Functions*, 2nd Edition (1895), Chap. VI.

For numerical calculation, form the A.G.M. scale ($a_0 = 1$, $b_0 = k'$), and starting with $\phi_0 = \theta$, make use of the recurrence formula (17),

$$\tan (\phi_{n+1} - \phi_n) = (b_n/a_n) \tan \phi_n.$$

We then have $a_n a = \phi_n/2^n$, and as in Case I, we derive finally

$$\Pi_3 (n, k, \phi)$$
$$= - u \tan \phi_0 \sec \phi_0 \cos (\phi_1 - \phi_0) \{c_1 \sin \phi_1 + c_2 \sin \phi_2 + \ldots + c_n \sin \phi_n + \ldots\}$$
$$- \tfrac{1}{2} \tan \phi_0 \sec \phi_0 \cos (\phi_1 - \phi_0) \log [H (a - u)/H (a + u)] \quad \ldots \ldots (110).$$

Starting with $\phi_0 = \phi$ and making use of the same recurrence formula as above, the computation of u presents no difficulty. As before, the calculation of the last term of (110) is best achieved by computing the nome q by (64) and making use of (108).

If we write $\phi = \tfrac{1}{2}\pi$, the last term of (110) vanishes, and we obtain for the complete elliptic integral of the third kind the expression

$$\Pi_3 (n, k, \tfrac{1}{2}\pi)$$
$$= - K \tan \phi_0 \sec \phi_0 \cos (\phi_1 - \phi_0) \{c_1 \sin \phi_1 + c_2 \sin \phi_2 + \ldots + c_n \sin \phi_n + \ldots\}$$
$$\ldots \ldots (111).$$

When $u > a$, it will be noticed that the integral (105) has a singularity at $u = a$, where the integrand becomes infinite. At the same time the logarithm of a negative quantity appears in (110), since $H (a - u) = - H (u - a)$. It is necessary in this instance to define the path of integration, regarding (105) as an algebraic integral obtained by writing $z = \operatorname{sn} u$. Near the point $z = \operatorname{sn} a$ this integral is approximately

$$\Pi_3 = \operatorname{sn}^2 a/(\operatorname{cn} a \operatorname{dn} a) \cdot \int dz/(\operatorname{sn}^2 a - z^2),$$

which integrated round a small semi-circle below the real axis gives rise to the imaginary term $- \tfrac{1}{2}\pi i \operatorname{sn} a/(\operatorname{cn} a \operatorname{dn} a)$. If we define the path of integration of the algebraic form of the integral as consisting of this small semi-circle and the remaining portions of the real axis, the above imaginary term cancels the term $- \operatorname{sn} a/(\operatorname{cn} a \operatorname{dn} a) \cdot \tfrac{1}{2} \log (- 1)$ appearing on the right-hand side of (110) when we write $H (a - u) = - H (u - a)$. With this meaning attached to the path of integration, we may write the incomplete integral in the form, $u > a$,

$$\Pi_3 (n, k, \phi) = - \operatorname{sn} a/(\operatorname{cn} a \cdot \operatorname{dn} a) [u Z (a, k) + \tfrac{1}{2} \log \{H (u - a)/H (u + a)\}]$$
$$\ldots \ldots (112).$$

Numerical computation proceeds in the manner already described.

CASE III. *n negative, between* $- k^2$ *and* $- 1$. *Circular case.*

First Method.

We write

$$n = - k^2 \operatorname{sn}^2 (ia + K, k) = - k^2/\operatorname{dn}^2 (a, k') = - k^2/(1 - k'^2 \sin^2 \theta) \quad \ldots (113),$$
whence
$$\sin \theta = \operatorname{sn} (a, k'), \qquad 0 < a < K'.$$

Writing $ia + K$ for a in (98), we find

$$\Pi_3\,(n,\,k,\,\phi) = u + i\,\mathrm{dn}\,(a,\,k')/[k'^2\,\mathrm{sn}\,(a,\,k')\,\mathrm{cn}\,(a,\,k')]\,.\,\Pi\,(u,\,ia+K)$$
$$\dots\dots(114),$$

while (97) gives

$$\Pi\,(u,\,ia+K) = uZ\,(ia+K) + \tfrac{1}{2}\log\,[\Theta\,(ia-u+K)/\Theta\,(ia+u+K)]$$
$$\dots\dots(115).$$

Making use of the addition formula (79) for $Z\,(u)$, we find after some reductions

$$iZ\,(ia+K) = iZ\,(ia,\,k) + k^2\,\mathrm{sn}\,(a,\,k')/[\mathrm{cn}\,(a,\,k')\,\mathrm{dn}\,(a,\,k')]\ \dots(116),$$

or, from (47), in terms of a real argument

$$iZ\,(ia+K) = Z\,(a,\,k') + \tfrac{1}{2}\,\pi a/(KK') - k'^2\,\mathrm{sn}\,(a,\,k')\,\mathrm{cn}\,(a,\,k')/\mathrm{dn}\,(a,\,k')$$
$$\dots\dots(117).$$

Introducing Jacobi's Theta-function $\Theta_1\,(z)$ defined by the relations

$$\Theta\,(z+K) = \Theta_1\,(z)\ \dots\dots\dots\dots\dots(118),$$

$$\Theta_1\,(z) = 1 + 2q\cos 2a_n z + 2q^4\cos 4a_n z + 2q^9\cos 6a_n z + \dots\ \dots(119),$$

we may write (114) in either of the forms

$$\Pi_3\,(n,\,k,\,\phi) = \frac{\mathrm{dn}\,(a,\,k')}{k'^2\,\mathrm{sn}\,(a,\,k')\,.\,\mathrm{cn}\,(a,\,k')}$$
$$\times\left[\frac{u\,\mathrm{sn}(a,\,k')\,\mathrm{cn}\,(a,\,k')}{\mathrm{dn}\,(a,\,k')} - \frac{u}{i}\,Z\,(ia,\,k) + \tfrac{1}{2}\,i\log\frac{\Theta_1\,(u-ia)}{\Theta_1\,(u+ia)}\right]\ \dots(120),$$

$$\Pi_3\,(n,\,k,\,\phi) = \frac{\mathrm{dn}\,(a,\,k')}{k'^2\,\mathrm{sn}\,(a,\,k')\,.\,\mathrm{cn}\,(a,\,k')}$$
$$\times\left[\frac{\pi au}{2KK'} + uZ\,(a,\,k') + \tfrac{1}{2}\,i\log\frac{\Theta_1\,(u-ia)}{\Theta_1\,(u+ia)}\right]\ \dots(121).$$

For numerical computation we may make use of two sets of formulae, according as we employ circular or hyperbolic functions.

(i) *Use of circular recurrence formulae.*

Form the A.G.M. scale $(a_0 = 1,\ b_0 = k')$. Commencing with $\phi_0 = \phi$, make use of the recurrence formula (17), $\tan\,(\phi_{n+1} - \phi_n) = (b_n/a_n)\tan\phi_n$, and thus derive $a_n u = \phi_n/2^n = \Phi$.

Commencing with $\psi_0 = \theta$, employ the recurrence formula (67), $\sin\,(2\psi_{n+1} - \psi_n) = (b_n/a_n)\sin\psi_n$, giving $a_n a = \log\tan\,(\tfrac{1}{4}\pi + \tfrac{1}{2}\psi_n)$. The application of (47) and (76) to both (120) and (121) gives us the formula

$$\Pi_3\,(n,\,k,\,\phi) = \frac{2\cos\,(2\psi_1 - \psi_0)}{k'^2\sin 2\psi_0}\left[u\,\{a_n\sin\psi_n + \sum_2(\psi)\} + \tfrac{1}{2}\,i\log\frac{\Theta_1\,(u-ia)}{\Theta_1\,(u+ia)}\right]$$
$$\dots\dots(122)$$

for the numerical calculation of the incomplete integral, and the formula

$$\Pi_3\,(n,\,k,\,\tfrac{1}{2}\pi) = 2\,(K/k'^2)\cos\,(2\psi_1 - \psi_0)/\sin 2\psi_0\,.\,[a_n\sin\psi_n + \sum_2(\psi)]$$
$$\dots\dots(123).$$

The peculiar difficulty arising in the numerical computation of the circular cases of the incomplete elliptic integral of the third kind lies in the evaluation of the Theta-functions of complex argument. Writing $z = u + ia$ in Jacobi's expansion (119), the series may be resolved into real and imaginary parts[*]. If we write

$$\tan \Psi = \frac{2q \sin 2a_n u \cdot \sinh 2a_n a + 2q^4 \sin 4a_n u \cdot \sinh 4a_n a + \ldots}{1 + 2q \cos 2a_n u \cdot \cosh 2a_n a + 2q^4 \cos 4a_n u \cdot \cosh 4a_n a + \ldots}$$
$$\ldots\ldots(124),$$

we easily find $\log [\Theta_1 (u - ia)/\Theta_1 (u + ia)] = 2i\Psi$, from which it follows that (122) may be written

$$\Pi_3 (n, k, \phi) = (2/k'^2) \cos (2\psi_1 - \psi_0)/\sin 2\psi_0 [u \{a_n \sin \psi_n + \sum_2 (\psi)\} - \Psi]$$
$$\ldots\ldots(125).$$

Powers of the nome q arising in (124) are most conveniently calculated from (64),

$$\log q = \sum_1^\infty (\tfrac{1}{2})^{n-1} \log (a_n/a_{n+1}) - \log (4a_1/c_1).$$

The use of hyperbolic functions in (124) may be avoided by noting that if we write

$$Q^2 = e^{2a_n a} = \tan^2 (\tfrac{1}{4}\pi + \tfrac{1}{2}\psi_n) = (1 + \sin \psi_n)/(1 - \sin \psi_n),$$

$$\left.\begin{array}{l} 2 \cosh 2a_n a = Q^2 + Q^{-2} \\ 2 \sinh 2a_n a = Q^2 - Q^{-2} \end{array}\right\}, \quad \left.\begin{array}{l} 2 \cosh 4a_n a = Q^4 + Q^{-4} \\ 2 \sinh 4a_n a = Q^4 - Q^{-4} \end{array}\right\}, \ldots \text{etc.},$$

from which it follows that (124) may be written

$$\tan \Psi = \frac{q(Q^2 - Q^{-2}) \sin 2\Phi + q^4(Q^4 - Q^{-4}) \sin 4\Phi + q^9(Q^6 - Q^{-6}) \sin 6\Phi + \ldots}{1 + q(Q^2 + Q^{-2})\cos 2\Phi + q^4(Q^4 + Q^{-4})\cos 4\Phi + q^9(Q^6 + Q^{-6})\cos 6\Phi + \ldots}$$
$$\ldots\ldots(124 \ bis).$$

(ii) *Use of hyperbolic recurrence formulae.*

Write $n = - k^2 \operatorname{sn}^2 (ia + K, k) = - k^2/(1 - k'^2 \tanh^2 \phi_0') \ldots\ldots(126)$,
whence $\operatorname{sn} (a, k') = \tanh \phi_0', \ 0 < a < K'$.

From the hyperbolic recurrence formula (37)

$$\tanh (\phi'_{n+1} - \phi_n') = (b_n/a_n) \tanh \phi_n'$$

we have

$$a_n a = \phi_n'/2^n = \Phi', \quad \operatorname{sn} (a, k') = \tanh \phi_0',$$
$$\operatorname{cn} (a, k') = \operatorname{sech} \phi_0', \quad \operatorname{dn} (a, k') = \operatorname{sech} (\phi_1' - \phi_0').$$

---

[*] This procedure is due to Legendre, *Fonctions Elliptiques*, Deuxième Supplement (1828), § IX. p. 146 *et seq.*

Thus from (44) and (120) we obtain

$$\Pi_3(n, k, \phi) = \frac{1}{k'^2} \frac{\cosh \phi_0'}{\tanh \phi_0' \cosh(\phi_1' - \phi_0')}$$
$$\times \left[\frac{u \sinh \phi_0'}{\cosh(\phi_1' - \phi_0')} - u \sum_1^\infty c_n \sinh \phi_n' + \tfrac{1}{2} i \log \frac{\Theta_1(u - ia)}{\Theta_1(u + ia)} \right] \quad \dots (127).$$

To calculate u we make use of the recurrence formula (17)

$$\tan(\phi_{n+1} - \phi_n) = (b_n/a_n) \tan \phi_n, \text{ giving } a_n u = \phi_n/2^n = \Phi.$$

Equation (124) then becomes

$$\tan \Psi = \frac{2q \sin 2\Phi \sinh 2\Phi' + 2q^4 \sin 4\Phi \sinh 4\Phi' + 2q^9 \sin 6\Phi \sinh 6\Phi' + \dots}{1 + 2q \cos 2\Phi \cosh 2\Phi' + 2q^4 \cos 4\Phi \cosh 4\Phi' + 2q^9 \cos 6\Phi \cosh 6\Phi' + \dots}$$
$$\dots\dots(128).$$

Equation (127) then gives

$$\Pi_3(n, k, \phi) = \frac{1}{k'^2} \frac{\cosh \phi_0'}{\tanh \phi_0' \cosh(\phi_1' - \phi_0')}$$
$$\times \left[\frac{u \sinh \phi_0'}{\cosh(\phi_1' - \phi_0')} - u \sum_1^\infty c_n \sinh \phi_n' - \Psi \right] \quad \dots\dots(129),$$

and

$$\Pi_3(n, k, \tfrac{1}{2}\pi) = \frac{K}{k'^2} \frac{\cosh \phi_0'}{\tanh \phi_0' \cosh(\phi_1' - \phi_0')} \left[\frac{\sinh \phi_0'}{\cosh(\phi_1' - \phi_0')} - \sum_1^\infty c_n \sinh \phi_n' \right]$$
$$\dots\dots(129 \text{ bis}).$$

Second Method.

If we write $\beta = K' - a$, (113) takes the form

$$n = -k^2 \operatorname{sn}^2(K + iK' - i\beta) = -\operatorname{dn}^2(\beta, k') = -1 + k'^2 \sin^2 \theta', \quad 0 < \beta < K'$$
$$\dots\dots(130).$$

We easily find

$$\Pi_3(n, k, \phi) = u + \frac{\operatorname{dn}(\beta, k')}{k'^2 \operatorname{sn}(\beta, k') \operatorname{cn}(\beta, k')}$$
$$\times \left[iu Z(K + iK' - i\beta) + \tfrac{1}{2} i \log \frac{\Theta(K + iK' - i\beta - u)}{\Theta(K + iK' - i\beta + u)} \right]$$
$$= u + \frac{\operatorname{dn}(\beta, k')}{k'^2 \operatorname{sn}(\beta, k') \operatorname{cn}(\beta, k')}$$
$$\times \left[iu Z(K + iK' - i\beta) - \frac{\pi u}{2K} + \tfrac{1}{2} i \log \frac{H_1(u + i\beta)}{H_1(u - i\beta)} \right] \quad \dots\dots(131)$$
$$= \frac{\operatorname{dn}(\beta, k')}{k'^2 \operatorname{sn}(\beta, k') \operatorname{cn}(\beta, k')}$$
$$\times \left[-\frac{uk^2 \operatorname{sn}(\beta, k')}{\operatorname{cn}(\beta, k') \operatorname{dn}(\beta, k')} - iu Z(i\beta) + \tfrac{1}{2} i \log \frac{H_1(u + i\beta)}{H_1(u - i\beta)} \right] \dots (132),$$

where $H_1(z)$ is Jacobi's Theta-function defined by the series

$$H_1(z) = 2q^{\frac{1}{4}}(\cos a_n z + q^2 \cos 3a_n z + q^6 \cos 5a_n z + q^{12} \cos 7a_n z + \dots)$$
$$\dots\dots(133).$$

Writing $\quad \Psi' = \tfrac{1}{2} i \log \left[H_1 (u + i\beta)/H_1 (u - i\beta) \right] \quad \ldots\ldots\ldots(134)$,

we find

$$\tan \Psi' = \frac{\sin a_n u \sinh a_n \beta + q^2 \sin 3 a_n u \sinh 3 a_n \beta + q^6 \sin 5 a_n u \sinh 5 a_n \beta + \ldots}{\cos a_n u \cosh a_n \beta + q^2 \cos 3 a_n u \cosh 3 a_n \beta + q^6 \cos 5 a_n u \cosh 5 a_n \beta + \ldots}$$
$$\ldots\ldots(135).$$

The numerical computation may be carried out by two methods.

(i) *Use of circular recurrence formulae.*

Starting with $\psi_0 = \theta'$ in (130), make use of the recurrence formula (67), $\sin (2\psi_{n+1} - \psi_n) = (b_n/a_n) \sin \psi_n$. From (80) we have

$$iZ (K + iK' - i\beta) = a_n (1 - \sin \psi_n) - \underset{2}{\Sigma} (\psi),$$

while

$$\operatorname{sn} (\beta, k') = \sin \psi_0, \quad \operatorname{cn} (\beta, k') = \cos \psi_0, \quad \operatorname{dn} (\beta, k') = \cos (2\psi_1 - \psi_0).$$

Thus (131) becomes

$$\Pi_3 (n, k, \phi) = u + \frac{2 \cos (2\psi_1 - \psi_0)}{k'^2 \sin 2\psi_0} \left[\Psi' - u \{ a_n \sin \psi_n + \underset{2}{\Sigma} (\psi) \} \right] \quad \ldots(136).$$

As before, u is most conveniently calculated from the recurrence formula $\tan (\phi_{n+1} - \phi_n) = (b_n/a_n) \tan \phi_n$, giving $a_n u = \phi_n/2^n = \Phi$.

The use of hyperbolic functions in (135) may be avoided by writing $\log Q' = a_n \beta = \log \tan (\tfrac{1}{4}\pi + \tfrac{1}{2}\psi_n)$, so that

$$\left. \begin{array}{l} 2 \cosh a_n \beta = Q' + Q'^{-1} \\ 2 \sinh a_n \beta = Q' - Q'^{-1} \end{array} \right\}, \quad \left. \begin{array}{l} 2 \cosh 3 a_n \beta = Q'^3 + Q'^{-3} \\ 2 \sinh 3 a_n \beta = Q'^3 - Q'^{-3} \end{array} \right\}, \quad \text{etc.} \quad \ldots(137).$$

Thus (135) may be written

$$\tan \Psi' = \frac{(Q' - Q'^{-1}) \sin \Phi + q^2 (Q'^3 - Q'^{-3}) \sin 3\Phi + q^6 (Q'^5 - Q'^{-5}) \sin 5\Phi + \ldots}{(Q' + Q'^{-1}) \cos \Phi + q^2 (Q'^3 + Q'^{-3}) \cos 3\Phi + q^6 (Q'^5 + Q'^{-5}) \cos 5\Phi + \ldots}$$
$$\ldots\ldots(138).$$

Writing $u = K$ in (136), $\Psi' = \tfrac{1}{2}\pi$, and we have, since $\tfrac{1}{2}\pi = a_n K$,

$$\Pi_3 (n, k, \tfrac{1}{2}\pi) = K \left[1 + \frac{2 \cos (2\psi_1 - \psi_0)}{k'^2 \sin 2\psi_0} \{ a_n (1 - \sin \psi_n) - \underset{2}{\Sigma} (\psi) \} \right] \ldots(139).$$

(ii) *Use of hyperbolic recurrence formulae.*

Write $\sin \theta' = \operatorname{sn} (\beta, k') = \tanh \phi_0'$, and (130) becomes

$$n = - k^2 \operatorname{sn}^2 (K + iK' - i\beta) = -1 + k'^2 \tanh^2 \phi_0' \quad \ldots(140).$$

Make use of the hyperbolic recurrence formula (37)

$$\tanh (\phi'_{n+1} - \phi_n') = (b_n/a_n) \tanh \phi_n',$$

and we have

$$a_n \beta = \Phi' = \phi_n'/2^n, \quad \operatorname{sn} (\beta, k') = \tanh \phi_0',$$
$$\operatorname{cn} (\beta, k') = \operatorname{sech} \phi_0', \quad \operatorname{dn} (\beta, k') = \operatorname{sech} (\phi_1' - \phi_0').$$

Also $\quad - iZ (i\beta, k) = c_1 \sinh \phi_1' + c_2 \sinh \phi_2' + \ldots + c_n \sinh \phi_n' + \ldots,$

so that (132) gives

$$\Pi_3 (n, k, \phi)$$

$$= \frac{1}{k'^2} \frac{\cosh \phi_0'}{\cosh (\phi_1' - \phi_0') \tanh \phi_0'}$$

$$\times \left[-uk^2 \sinh \phi_0' \cosh (\phi_1' - \phi_0') + u \sum_1^\infty c_n \sinh \phi_n' + \Psi' \right] \dots (141).$$

As before, we calculate u from the recurrence formula

$$\tan (\phi_{n+1} - \phi_n) = (b_n/a_n) \tan \phi_n,$$

giving $a_n u = \phi_n/2^n = \Phi$. Equation (135) gives for Ψ'

$$\tan \Psi' = \frac{\sinh \Phi' \sin \Phi + q^2 \sinh 3\Phi' \sin 3\Phi + q^6 \sinh 5\Phi' \sin 5\Phi + \dots}{\cosh \Phi' \cos \Phi + q^2 \cosh 3\Phi' \cos 3\Phi + q^6 \cosh 5\Phi' \cos 5\Phi + \dots}$$

$$\dots \dots (142).$$

Writing $u = K$ in (141), we have $\Psi' = \frac{1}{2}\pi$, and since $\frac{1}{2}\pi = a_n K$,

$$\Pi_3 (n, k, \tfrac{1}{2}\pi)$$

$$= \frac{K \cosh \phi_0'}{k'^2 \cosh (\phi_1' - \phi_0') \tanh \phi_0'} \left[a_n + \sum_0^\infty c_n \sinh \phi_n' - k^2 \sinh \phi_0' \cosh (\phi_1' - \phi_0') \right]$$

$$\dots \dots (143).$$

CASE IV. *n positive, between* 0 *and* ∞. *Circular case.*

First Method.

Write $\quad n = -k^2 \operatorname{sn}^2 (ia, k) = k^2 \operatorname{sn}^2 (a, k')/\operatorname{cn}^2 (a, k') = k^2 \tan^2 \theta \dots (144),$

giving $\qquad\qquad \sin \theta = \operatorname{sn} (a, k'), \quad 0 < a < K'.$

Thus

$$\Pi_3 (n, k, \phi) = u + \frac{\operatorname{sn} (a, k') \operatorname{cn} (a, k')}{\operatorname{dn} (a, k')} \left[iu Z(ia) + \tfrac{1}{2}i \log \frac{\Theta (u - ia)}{\Theta (u + ia)} \right]$$

$$\dots \dots (145).$$

$\Theta (z)$ is Jacobi's Theta-function defined by the series

$$\Theta (z) = 1 - 2q \cos 2a_n z + 2q^4 \cos 4a_n z - 2q^9 \cos 6a_n z + \dots (146).$$

Writing $\qquad\qquad \Psi = \tfrac{1}{2}i \log [\Theta (u - ia)/\Theta (u + ia)] \quad \dots \dots \dots (147),$

we find

$$\tan \Psi = \frac{2q \sin 2a_n u \sinh 2a_n a - 2q^4 \sin 4a_n u \sinh 4a_n a + \dots}{1 - 2q \cos 2a_n u \cosh 2a_n a + 2q^4 \cos 4a_n u \cosh 4a_n a - \dots}$$

$$\dots \dots (148).$$

The numerical computation may be carried out by two methods.

(i) *Use of circular recurrence formulae.*

Make use of the recurrence formula $\sin (2\psi_{n+1} - \psi_n) = (b_n/a_n) \sin \psi_n$, commencing with $\psi_0 = \theta$ in (144). We then have

$$\operatorname{sn} (a, k') = \sin \psi_0, \quad \operatorname{cn} (a, k') = \cos \psi_0, \quad \operatorname{dn} (a, k') = \cos (2\psi_1 - \psi_0),$$

and by (78)

$$iZ (ia, k) = a_n \sin \psi_n + \sum_2 (\psi) - \operatorname{sn} (a, k') \operatorname{dn} (a, k')/\operatorname{cn} (a, k') \dots (149).$$

Hence (145) may be written

$$\Pi_3\,(n, k, \phi) = \frac{1}{2}\,\frac{\sin 2\psi_0}{\cos(2\psi_1 - \psi_0)}\left[\,u\left\{\frac{\cos(2\psi_1 - \psi_0)}{\tan \psi_0} + a_n \sin\psi_n + \underset{2}{\Sigma}\,(\psi)\right\} + \Psi\,\right]$$

$$\dots\dots(150).$$

As before, u is calculated from the recurrence formula

$$\tan(\phi_{n+1} - \phi_n) = (b_n/a_n)\tan\phi_n,$$

giving $a_n u = \phi_n/2^n = \Phi$, while $a_n a = \log\tan(\tfrac{1}{4}\pi + \tfrac{1}{2}\psi_n)$. We have

$$\begin{array}{ll}
2\cosh 2a_n a = Q^2 + Q^{-2} \\
2\sinh 2a_n a = Q^2 - Q^{-2}
\end{array}\Bigg\}\,,\quad
\begin{array}{ll}
2\cosh 4a_n a = Q^4 + Q^{-4} \\
2\sinh 4a_n a = Q^4 - Q^{-4}
\end{array}\Bigg\}\,,\text{ etc. }\dots\dots(151),$$

where

$$Q = \tan(\tfrac{1}{4}\pi + \psi_n),$$

so that (148) becomes

$$\tan\Psi = \frac{q\,(Q^2 - Q^{-2})\sin 2\Phi - q^4\,(Q^4 - Q^{-4})\sin 4\Phi + q^9\,(Q^6 - Q^{-6})\sin 6\Phi - \dots}{1 - q\,(Q^2 + Q^{-2})\cos 2\Phi + q^4\,(Q^4 + Q^{-4})\cos 4\Phi - q^9\,(Q^6 + Q^{-6})\cos 6\Phi - \dots}$$

$$\dots\dots(152).$$

Writing $u = K$ in (150), $\Psi = 0$, and

$$\Pi_3\,(n, k, \tfrac{1}{2}\pi) = \frac{K\sin 2\psi_0}{2\cos(2\psi_1 - \psi_0)}\left[\frac{\cos(2\psi_1 - \psi_0)}{\tan\psi_0} + a_n \sin\psi_n + \underset{2}{\Sigma}\,(\psi)\right]$$

$$\dots\dots(153).$$

(ii) *Use of hyperbolic recurrence formulae.*

Write

$$\sin\theta = \operatorname{sn}(a, k') = \tanh\phi_0' \dots\dots\dots\dots\dots(154),$$

so that

$$n = -k^2\operatorname{sn}^2(ia, k) = k^2\tan^2\theta = k^2\sinh^2\phi_0'.$$

Starting with ϕ_0' given by (154), make use of the recurrence formula

$$\tanh(\phi'_{n+1} - \phi_n') = (b_n/a_n)\tanh\phi_n' \dots\dots\dots(155).$$

We then have

$$\begin{aligned}
a_n a &= \phi_n'/2^n = \Phi', & \operatorname{sn}(a, k') &= \tanh\phi_0', \\
\operatorname{cn}(a, k') &= \operatorname{sech}\phi_0', & \operatorname{dn}(a, k') &= \operatorname{sech}(\phi_1' - \phi_0'),
\end{aligned}$$

and

$$-iZ(ia, k) = \sum_1^\infty c_n \sinh\phi_n'.$$

Equation (145) may then be written

$$\Pi_3\,(n, k, \phi) = u + \frac{\tanh\phi_0'\cosh(\phi_1' - \phi_0')}{\cosh\phi_0'}\left[\Psi - u\sum_1^\infty c_n \sinh\phi_n'\right]\dots(156).$$

As before, u is calculated from the recurrence formula

$$\tan(\phi_{n+1} - \phi_n) = (b_n/a_n)\tan\phi_n,$$

giving $a_n u = \phi_n/2^n = \Phi$. Equation (148) then gives for Ψ

$$\tan\Psi = \frac{2q\sinh 2\Phi'\sin 2\Phi - 2q^4\sinh 4\Phi'\sin 4\Phi + 2q^9\sinh 6\Phi'\sin 6\Phi - \dots}{1 - 2q\cosh 2\Phi'\cos 2\Phi + 2q^4\cosh 4\Phi'\cos 4\Phi - 2q^9\cosh 6\Phi'\cos 6\Phi + \dots}$$

$$\dots\dots(157).$$

Writing $u = K$ in (156), $\Psi = 0$, and

$$\Pi_3\left(n, k, \tfrac{1}{2}\pi\right) = K\left[1 - \frac{\tanh \phi_0' \cosh(\phi_1' - \phi_0')}{\cosh \phi_0'} \sum_1^\infty c_n \sinh \phi_n'\right] \quad \ldots(158).$$

Second Method.

If we write $\beta = K' - \alpha$ in (145), we have

$$n = -k^2 \operatorname{sn}^2(iK' - i\beta) = \operatorname{cn}^2(\beta, k')/\operatorname{sn}^2(\beta, k') = \cot^2 \theta \quad \ldots(159),$$

giving $\quad \operatorname{sn}(\beta, k') = \sin\theta, \quad 0 < \beta < K'$.

We easily find

$\Pi_3(n, k, \phi)$

$$= u + \frac{\operatorname{sn}(\beta, k')\operatorname{cn}(\beta, k')}{\operatorname{dn}(\beta, k')}\left[iuZ(iK' - i\beta) + \tfrac{1}{2}i\log\frac{\Theta(iK' - i\beta - u)}{\Theta(iK' - i\beta + u)}\right]$$

$$= \frac{\operatorname{sn}(\beta, k')\operatorname{cn}(\beta, k')}{\operatorname{dn}(\beta, k')}\left[-iuZ(i\beta) + \tfrac{1}{2}i\log\frac{H(u + i\beta)}{H(-u + i\beta)}\right] \quad \ldots\ldots(160),$$

where $H(z)$ is Jacobi's Theta-function defined by

$$H(z) = 2q^{\frac{1}{4}}(\sin a_n z - q^2 \sin 3a_n z + q^6 \sin 5a_n z - q^{12} \sin 7a_n z + \ldots)$$
$$\ldots\ldots(161).$$

If we write

$$\Psi' = \tfrac{1}{2}i\log[H(u + i\beta)/H(-u + i\beta)] \quad \ldots\ldots(162),$$

we find

$$\cot\Psi' = \frac{\cos a_n u \sinh a_n\beta - q^2\cos 3a_n u \sinh 3a_n\beta + q^6\cos 5a_n u \sinh 5a_n\beta - \ldots}{\sin a_n u \cosh a_n\beta - q^2\sin 3a_n u \cosh 3a_n\beta + q^6\sin 5a_n u \cosh 5a_n\beta - \ldots}$$
$$\ldots\ldots(163).$$

The numerical computation can be carried out by two methods.

(i) *Use of circular recurrence formulae.*

Commencing with $\psi_0 = \theta$ in (159), make use of the recurrence formula (67)

$$\sin(2\psi_{n+1} - \psi_n) = (b_n/a_n)\sin\psi_n.$$

From (78) we have

$$-iZ(i\beta, k) = \tan\psi_0\cos(2\psi_1 - \psi_0) - a_n\sin\psi_n - \sum_2(\psi) \quad \ldots(164),$$

while $\operatorname{sn}(\beta, k') = \sin\psi_0$, $\operatorname{cn}(\beta, k') = \cos\psi_0$, $\operatorname{dn}(\beta, k') = \cos(2\psi_1 - \psi_0)$.

Equation (160) then gives

$\Pi_3(n, k, \phi)$

$$= \frac{\sin 2\psi_0}{2\cos(2\psi_1 - \psi_0)}\left[u\left\{\tan\psi_0\cos(2\psi_1 - \psi_0) - a_n\sin\psi_n - \sum_2(\psi)\right\} + \Psi'\right]$$
$$\ldots\ldots(165).$$

As before, u is calculated from the recurrence formula

$$\tan(\phi_{n+1} - \phi_n) = (b_n/a_n)\tan\phi_n,$$

giving $a_n u = \phi_n/2^n = \Phi$, while $a_n \beta = \log \tan\left(\frac{1}{4}\pi + \frac{1}{2}\psi_n\right)$. Thus

$$\left.\begin{array}{ll} 2\cosh a_n\beta = Q' + Q'^{-1} \\ 2\sinh a_n\beta = Q' - Q'^{-1} \end{array}\right\}, \quad \left.\begin{array}{ll} 2\cosh 3a_n\beta = Q'^3 + Q'^{-3} \\ 2\sinh 3a_n\beta = Q'^3 - Q'^{-3} \end{array}\right\}, \text{ etc.} \quad \ldots\ldots(166),$$

where $\qquad\qquad Q' = \tan\left(\frac{1}{4}\pi + \frac{1}{2}\psi_n\right)$.

Thus (163) becomes

$$\tan \Psi' = \frac{(Q' - Q'^{-1})\cos\Phi - q^2(Q'^3 - Q'^{-3})\cos 3\Phi + q^6(Q'^5 - Q'^{-5})\cos 5\Phi - \ldots}{(Q' + Q'^{-1})\sin\Phi - q^2(Q'^3 + Q'^{-3})\sin 3\Phi + q^6(Q'^5 + Q'^{-5})\sin 5\Phi - \ldots}$$

$$\ldots\ldots(167).$$

Writing $u = K$ in (163), $\Psi' = \frac{1}{2}\pi$, and (165) becomes

$$\Pi_3\left(n, k, \tfrac{1}{2}\pi\right) = \frac{K\sin 2\psi_0}{2\cos(2\psi_1 - \psi_0)}\left[\tan\psi_0\cos(2\psi_1 - \psi_0) + a_n(1 - \sin\psi_n) - \underset{2}{\Sigma}(\psi)\right]$$

$$\ldots\ldots(168).$$

(ii) *Use of hyperbolic recurrence formulae.*

Write $\sin\theta' = \text{sn}\,(\beta, k') = \tanh\phi_0'$, so that (159) gives

$$n = -k^2\,\text{sn}^2(iK' - i\beta) = \cot^2\theta' = 1/\sinh^2\phi_0', \quad 0 < \beta < K' \ldots(169).$$

Commencing with ϕ_0' in (169), use the recurrence formula

$$\tanh(\phi'_{n+1} - \phi_n') = (b_n/a_n)\tanh\phi_n' \quad\ldots\ldots\ldots\ldots(170).$$

We then have

$$a_n\beta = \phi_n'/2^n = \Phi', \qquad \text{sn}\,(\beta, k') = \tanh\phi_0',$$
$$\text{cn}\,(\beta, k') = \text{sech}\,\phi_0', \qquad \text{dn}\,(\beta, k') = \text{sech}\,(\phi_1' - \phi_0'),$$

and $\qquad\qquad -iZ(i\beta, k) = \sum_1^\infty c_n\sinh\phi_n'.$

Equation (160) then gives

$$\Pi_3(n, k, \phi) = \frac{\tanh\phi_0'\cosh(\phi_1' - \phi_0')}{\cosh\phi_0'}\left[u\sum_1^\infty c_n\sinh\phi_n' + \Psi'\right] \quad\ldots(171).$$

As before, u is calculated from the recurrence formula

$$\tan(\phi_{n+1} - \phi_n) = (b_n/a_n)\tan\phi_n,$$

giving $a_n u = \phi_n/2^n = \Phi$. Equation (163) then gives for Ψ' the expression

$$\cot\Psi' = \frac{\sinh\Phi'\cos\Phi - q^2\sinh 3\Phi'\cos 3\Phi + q^6\sinh 5\Phi'\cos 5\Phi - \ldots}{\cosh\Phi'\sin\Phi - q^2\cosh 3\Phi'\sin 3\Phi + q^6\cosh 5\Phi'\sin 5\Phi - \ldots}$$

$$\ldots\ldots(172).$$

Writing $u = K$ in (163), $\Psi' = \frac{1}{2}\pi$, and (171) becomes

$$\Pi_3\left(n, k, \tfrac{1}{2}\pi\right) = K\frac{\tanh\phi_0'\cosh(\phi_1' - \phi_0')}{\cosh\phi_0'}\left[a_n + \sum_1^\infty c_n\sinh\phi_n'\right]\ldots(173).$$

SECTION X

NOTE ON THE CALCULATION OF THE THIRD ELLIPTIC INTEGRAL IN TERMS OF THE COMPLEMENTARY A.G.M. SCALE

When k is very nearly unity, it may in some cases be advantageous to make use of the more convergent complementary scale ($a_0' = 1$, $b_0' = k$). The problem is to express the various functions appearing in the formula

$$\Pi_3(n, k, \phi) = u + \frac{\text{sn}(a, k)}{\text{cn}(a, k)\,\text{dn}(a, k)}\left[u\,Z(a, k) + \tfrac{1}{2}\log\frac{\Theta(a-u, k)}{\Theta(a+u, k)}\right]$$

$$\ldots\ldots(174)$$

in terms of recurrence formulae involving the complementary A.G.M. scale. The four cases discussed in Section IX, depending on the values of n, require separate treatment. The reader will have no difficulty in adapting the formulae of Section IX to the complementary scale should the occasion arise. We may note, in passing, the formula

$$\frac{\Theta(a-u)}{\Theta(a+u)} = e^{\frac{\pi a u}{KK'}}\frac{\text{cn}(a+u)}{\text{cn}(a-u)}\cdot\frac{\Theta\{i(a-u), k'\}}{\Theta\{i(a+u), k'\}}\ldots\ldots(175)$$

which will be found to be useful.

SUMMARY OF FORMULAE

NUMERICAL CALCULATION OF ELLIPTIC FUNCTIONS AND INTEGRALS

[Sections containing new formulae are denoted by an asterisk *.]

First Elliptic Integral. $u = F(\phi, k) = \int_0^\phi \dfrac{d\phi}{\sqrt{(1 - k^2 \sin^2 \phi)}}$, $K = F(\tfrac{1}{2}\pi, k)$.

Second Elliptic Integral. $E(\phi, k) = \int_0^\phi \sqrt{(1 - k^2 \sin^2 \phi)}\, d\phi$, $E = E(\tfrac{1}{2}\pi, k)$.

Third Elliptic Integral. $\Pi_3(n, k, \phi) = \int_0^\phi \dfrac{d\phi}{(1 + n \sin^2 \phi)\sqrt{(1 - k^2 \sin^2 \phi)}}$.

Jacobian Functions.

$$\sin \phi = \operatorname{sn}(u, k), \qquad \cos \phi = \operatorname{cn}(u, k), \qquad \sqrt{(1 - k^2 \sin^2 \phi)} = \operatorname{dn}(u, k).$$

Scale of Arithmetico-Geometrical Means.

$a_0 = 1$	$b_0 = k'$	$c_0 = k$
$a_1 = \tfrac{1}{2}(a_0 + b_0)$	$b_1 = \sqrt{a_0 b_0}$	$c_1 = \tfrac{1}{2}(a_0 - b_0)$
$a_2 = \tfrac{1}{2}(a_1 + b_1)$	$b_2 = \sqrt{a_1 b_1}$	$c_2 = \tfrac{1}{2}(a_1 - b_1)$
...	...	...
a_n	b_n	c_n

1. Landen's Scale of Increasing Amplitudes

$$\tan(\phi_{n+1} - \phi_n) = (b_n/a_n)\tan \phi_n.$$

$$F(\phi, k) = \operatorname*{Lt}_{n \to \infty}\left[\frac{1}{a_n}\frac{\phi_n}{2^n}\right], \qquad K = \tfrac{1}{2}\pi/a_n,$$

$$(K - E)/K = \tfrac{1}{2}(c_0^2 + 2c_1^2 + 4c_2^2 + \ldots + 2^n c_n^2 + \ldots),$$

$$Z(\phi, k) = E(\phi, k) - (E/K)F(\phi, k)$$
$$= c_1 \sin \phi_1 + c_2 \sin \phi_2 + \ldots + c_n \sin \phi_n + \ldots.$$

*2. *Calculation of* sn (u, k), cn (u, k), dn (u, k) *in terms of argument u.*
Calculate $\phi_n = 2^n a_n u$, n being chosen so that c_n is negligible. Obtain
$\phi_{n-1} \dots \phi_2$, ϕ_1, ϕ_0 from recurrence formula $\sin (2\phi_{n-1} - \phi_n) = \dfrac{c_n}{a_n} \sin \phi_n$.
Then sn $(u, k) = \sin \phi_0$, cn $(u, k) = \cos \phi_0$, dn $(u, k) = \cos \phi_0 \sec (\phi_1 - \phi_0)$.

*3. *The Hyperbolic Scale of Increasing Amplitudes*
$\tanh (\phi_{n+1} - \phi_n) = (b_n/a_n) \tanh \phi_n$. Start with $\sinh \phi_0 = \tan \phi$.
$$F(\phi, k') = \operatorname*{Lt}_{n \to \infty} [1/a_n \cdot \phi_n/2^n],$$

$$E(\phi, k') = \frac{\sinh \phi_0}{\cosh (\phi_1 - \phi_0)} + \frac{K - E}{K} F(\phi, k')$$
$$- (c_1 \sinh \phi_1 + c_2 \sinh \phi_2 + \dots + c_n \sinh \phi_n + \dots).$$

*4. *Calculation of* sn (u, k'), cn (u, k'), dn (u, k') *in terms of argument u.*
$$(1/i) \operatorname{sn} (iu, k) = \operatorname{sn} (u, k')/\operatorname{cn} (u, k'), \qquad \operatorname{cn} (iu, k) = 1/\operatorname{cn} (u, k'),$$
$$\operatorname{dn} (iu, k) = \operatorname{dn} (u, k')/\operatorname{cn} (u, k').$$

Calculate $\phi_n = 2^n a_n u$, n being chosen so that c_n is negligible. Obtain
$\phi_{n-1} \dots \phi_2$, ϕ_1, ϕ_0 from recurrence formula
$$\sinh (2\phi_{n-1} - \phi_n) = (c_n/a_n) \sinh \phi_n.$$
Then sn $(u, k') = \tanh \phi_0$, cn $(u, k') = \operatorname{sech} \phi_0$, dn $(u, k') = \operatorname{sech} (\phi_1 - \phi_0)$.

5. *Calculation of K', E', and the nome $q = e^{-\pi K'/K}$.*
**Hyperbolic recurrence formula* $\tanh \chi_{n+1} = \sqrt{(a_{n+1}/b_{n+1})} \tanh 2\chi_n$.
$$F(\phi, k') = \operatorname*{Lt}_{n \to \infty} [1/a_n \cdot \chi_n/2^n], \quad \tanh \chi_0 = \sqrt{k'} \sin \phi.$$

Put $\phi = \tfrac{1}{2}\pi$. $\tfrac{1}{2}\pi K'/K = \tfrac{1}{2}\log(1/q) = \operatorname{Lt}\left[(\tfrac{1}{2})^n \log \dfrac{a_n}{c_n} \right]_{n \to \infty}$. (Legendre.)

$$\log q = \sum_1^\infty (\tfrac{1}{2})^{n-1} \log \frac{a_n}{a_{n+1}} - \log \frac{4a_1}{c_1}. \qquad \text{(Gauss.)}$$

$$K' - E' = \tfrac{1}{2} \frac{1}{a_n} (1 - \tfrac{1}{2}c_0^2 - c_1^2 - 2c_2^2 - \dots - 2^{n-1}c_n^2 - \dots)$$
$$\times \left\{ \log \frac{4a_1}{c_1} - \log \frac{a_1}{a_2} - \tfrac{1}{2} \log \frac{a_2}{a_3} - \dots \right\} - a_n.$$

6. *Landen's Scale of Decreasing Amplitudes*
$$\sin (2\psi_{n+1} - \psi_n) = \frac{b_n}{a_n} \sin \psi_n.$$
$$F(\psi, k') = \frac{1}{a_n} \log \tan (\tfrac{1}{4}\pi + \tfrac{1}{2}\psi_n), \qquad \psi_0 = \psi.$$

$$*E\left(\psi, k'\right) = a_n \sin \psi_n + \tfrac{1}{2}\left(c_0{}^2 + 2c_1{}^2 + \ldots + 2^n c_n{}^2 + \ldots\right)$$

$$\times \frac{1}{a_n} \log \tan \left(\tfrac{1}{4}\pi + \tfrac{1}{2}\psi_n\right) + \underset{2}{\Sigma} \left(\psi_n, c_n\right),$$

$$\underset{2}{\Sigma} \left(\psi_n, c_n\right) = 2c_2 \frac{\tan \left(2\psi_2 - \psi_1\right)}{\cos \left(2\psi_3 - \psi_2\right)} + 6c_3 \frac{\tan \left(2\psi_3 - \psi_2\right)}{\cos \left(2\psi_4 - \psi_3\right)} + \ldots$$

$$+ 2\left(2^n - 1\right) c_{n+2} \frac{\tan \left(2\psi_{n+2} - \psi_{n+1}\right)}{\cos \left(2\psi_{n+3} - \psi_{n+2}\right)} + \ldots.$$

*7. Calculation of sn (u, k'), cn (u, k'), dn (u, k') **in terms of argument** u.

Calculate ψ_n from $\sin \psi_n = \tanh \left(a_n u\right)$, n being such that c_n is negligible. Obtain $\psi_{n-1} \ldots \psi_2$, ψ_1, ψ_0 from $\tan \left(\psi_{n-1} - \psi_n\right) = \frac{c_n}{a_n} \tan \psi_n$. Then

$$\text{sn} \left(u, k'\right) = \sin \psi_0, \quad \text{cn} \left(u, k'\right) = \cos \psi_0, \quad \text{dn} \left(u, k'\right) = \cos \left(2\psi_1 - \psi_0\right).$$

*8. Hyperbolic Scale of Decreasing Amplitudes

$$\sinh \left(2\psi_{n+1} - \psi_n\right) = \frac{b_n}{a_n} \sinh \psi_n.$$

$$F(\psi, k) = 1/a_n \tan^{-1} \left(\sinh \psi_n\right), \qquad \tan \psi = \sinh \psi_0,$$

$$E(\psi, k) - \frac{E}{K} F(\psi, k) = \tanh \psi_0 \cosh \left(2\psi_1 - \psi_0\right) - a_n \sinh \psi_n - \underset{2}{\Sigma'} \left(\psi_n, c_n\right).$$

*9. Calculation of sn (u, k), cn (u, k), dn (u, k).

Calculate ψ_n from $\sinh \psi_n = \tan a_n u$, n being such that c_n is negligible. Obtain $\psi_{n-1} \ldots \psi_2$, ψ_1, ψ_0 from $\tanh \left(\psi_{n-1} - \psi_n\right) = \left(c_n/a_n\right) \tanh \psi_n$. Then sn $(u, k) = \tanh \psi_0$, cn $(u, k) = \text{sech } \psi_0$, dn $(u, k) = \cosh \left(2\psi_1 - \psi_0\right)/\cosh \psi_0$.

*10. Calculation of Third Elliptic Integral—4 cases.

$$\Pi_3 \left(n, k, \phi\right) = u + \frac{\text{sn} \left(a, k\right)}{\text{cn} \left(a, k\right) \text{dn} \left(a, k\right)} \left[u Z\left(a, k\right) + \tfrac{1}{2} \log \frac{\Theta \left(a - u, k\right)}{\Theta \left(a + u, k\right)}\right].$$

Case I. $-k^2 < n < 0.$ Case II. $-\infty < k^2 < -1.$ Hyperbolic Cases.

Case III. $-1 < n < -k^2.$ ⎧Method 1. (i) Circular. (ii) Hyperbolic

⎪   recurrence formulae.

Case IV. $0 < n < +\infty.$ ⎨Method 2. (i) Circular. (ii) Hyperbolic

⎪   recurrence formulae.

⎩

Typical Solution. Case III, Method 1 (i).

$r = -k^2/\text{dn}^2 \left(a, k'\right) = -k^2/\left(1 - k'^2 \sin^2 \theta\right), \quad \sin \theta = \text{sn} \left(a, k'\right) \quad 0 < a < K'.$

1. Form A.G.M. scale $\left(a_0 = 1, b_0 = k'\right).$

2. Derive $a_n u = \phi_n/2^n = \Phi$ from $\tan \left(\phi_{n+1} - \phi_n\right) = \frac{b_n}{a_n} \tan \phi_n$, commencing with $\phi_0 = \phi.$

K   3

3. Derive $a_n a = \log \tan (\frac{1}{4}\pi + \frac{1}{2}\psi_n)$ from $\sin (2\psi_{n+1} - \psi_n) = \dfrac{b_n}{a_n} \sin \psi_n$, commencing with $\psi_0 = \theta$.

4. Compute q from $\log q = \overset{\infty}{\underset{1}{\Sigma}} (\frac{1}{2})^{n-1} \log (a_n/a_{n+1}) - \log (4a_1/c_1)$.

5. Compute Ψ from

$$\tan \Psi = \frac{2q \sin 2a_n u \sinh 2a_n a + 2q^4 \sin 4a_n u \sinh 4a_n a + 2q^9 + \ldots}{1 + 2q \cos 2a_n u \cosh 2a_n a + 2q^4 \cos 4a_n u \cosh 4a_n a + 2q^9 + \ldots}.$$

(Legendre.)

6. Then,

$$\Pi_3 (n, k, \phi) = \frac{2 \cos (2\psi_1 - \psi_0)}{k'^2 \sin 2\psi_0} [u \{a_n \sin \psi_n + \underset{2}{\Sigma} (\psi_n, c_n) - \Psi],$$

$$\Pi_3 (n, k, \tfrac{1}{2}\pi) = \frac{2K \cos (2\psi_1 - \psi_0)}{k'^2 \sin 2\psi_0} [a_n \sin \psi_n + \underset{2}{\Sigma} (\psi_n, c_n)].$$

APPENDIX

In the preceding sections only those formulae suitable for numerical computation have been given. Many interesting and elegant properties of the arithmetico-geometrical scales in conjunction with recurrence formulae not specially adapted to numerical calculation are given below as problems. In some cases the main steps of the proof are briefly indicated, and unless otherwise stated the final results are new.

Properties of the A.G.M. *scales* $(a_0 = 1, \ b_0 = k')$.

1. Prove that $a_n = a_{n+1} + c_{n+1}$, $b_n = a_{n+1} - c_{n+1}$. Hence derive the series

$$a_0 = a_n + c_1 + c_2 + c_3 + \dots,$$
$$b_0 = a_n - c_1 + c_2 + c_3 + \dots. \qquad \text{Gauss, } G. \ W. \text{ Bd. III. p. 376.}$$

2. Prove that

$$\sqrt{c_{n+2}} = \tfrac{1}{2}(\sqrt{a_n} - \sqrt{b_n}), \quad \sqrt{a_{n+2}} = \tfrac{1}{2}(\sqrt{a_n} + \sqrt{b_n}), \quad \sqrt{c_{n+2}} = \sqrt{a_n} - \sqrt{a_{n+2}}$$

and hence
$$\sqrt{a_n} = \sqrt{a_0} - \sqrt{c_2} - \sqrt{c_4} - \dots,$$
$$\sqrt{a_n} = \sqrt{b_0} + \sqrt{c_2} - \sqrt{c_4} - \dots. \qquad \text{Gauss, } G. \ W. \text{ Bd. III. p. 376.}$$

3. Prove that $c^2_n + 2c^2_{n+1} = 2(a_n + a_{n+1})c_{n+1}$, and hence derive the expansion

$$(K - E)/K = (a_0 + a_1)c_1 + 4(a_2 + a_3)c_3$$
$$+ 16(a_4 + a_5)c_5 + \dots + 2^{2n}(a_{2n} + a_{2n+1})c_{2n+1} + \dots.$$

4. Prove that $2b_{n-1}c_n = \tfrac{1}{2}c^2_{n-1} - 2c^2_n$, and hence derive the series

$$c_0{}^2 = 4(b_0 c_1 + 4b_1 c_2 + 16 b_2 c_3 + \dots + 2^{2n} b_n c_{n+1} + \dots),$$
$$a_n{}^2 = b_0{}^2 + 2(b_0 c_1 + b_1 c_2 + \dots + b_n c_{n+1} + \dots).$$

5. From the results of Ex. 4, prove that

$$E/K = b_0{}^2 + 2(b_0 c_1 + 2b_1 c_2 + 4b_2 c_3 + \dots + 2^{n-1} b_{n-1} c_n + \dots),$$
$$E/K = a_n{}^2 + 2(b_1 c_2 + 3b_2 c_3 + \dots + (2^{n-1} - 1)b_{n-1} c_n + \dots),$$

and hence the identity

$$1 = a_n{}^2 + 2\sum_2^\infty (2^{n-1} - 1)b_{n-1} c_n + \tfrac{1}{2}\sum_0^\infty 2^n c_n{}^2.$$

6. From the recurrence formula of Ex. 4 prove that

$$2(b_0 c_1 + b_1 c_2 + \dots) = \tfrac{1}{2}c_0{}^2 - \tfrac{3}{2}(c_1{}^2 + c_2{}^2 + c_3{}^2 + \dots),$$

and hence
$$a_n{}^2 = b_0{}^2 + \tfrac{1}{2}c_0{}^2 - \tfrac{3}{2}(c_1{}^2 + c_2{}^2 + c_3{}^2 + \dots),$$
$$a_n{}^2 = a_0{}^2 - \tfrac{1}{2}c_0{}^2 - \tfrac{3}{2}(c_1{}^2 + c_2{}^2 + c_3{}^2 + \dots).$$

$$\text{Gauss, } G. \ W. \text{ Bd. III. p. 376.}$$

7. Derive the recurrence formula

$$\log(a_{n-1}/a_n) = \log(b_n/b_{n-1}) - 2\log(b_{n+1}/b_n)$$

and hence prove that

$$\tfrac{1}{2}\pi K'/K = \log (4a_0/c_0) - \log (a_0/a_1) - \tfrac{1}{2}\log (a_1/a_2) - \dots$$
$$- (1/2^n)\log (a_n/a_{n+1}) - \dots,$$
$$\tfrac{1}{2}\pi K'/K = \log (4b_1/c_0) + \tfrac{3}{2}\log (b_2/b_1) + \tfrac{3}{4}\log (b_3/b_2) + \dots$$
$$+ (3/2^n)\log (b_{n+1}/b_n) + \dots.$$

Gauss, G. W. Bd. III. p. 377.

[Note that $\tfrac{1}{2}\pi K'/K = (1/2^n)\log (a_n/c_n)$ as n increases indefinitely.]

8. Derive the recurrence formula

$$(1/2^{n-1})\log (a_n/a_{n-1}) = (3/2^{n+1})\log (a_n/b_n)$$
$$+ (1/2^{n+1})\log (a_n/b_n) - (1/2^n)\log (a_{n-1}/b_{n-1}).$$

Hence prove that

$$\pi K'/K = \log (16a_0 b_0/c_0{}^2) + 3\sum_1^\infty (1/2^n)\log (a_n/b_n).$$

Jacobi, F. N. § 52, p. 201.

9. Show that the A.G.M. scale $(a_0 = 1,\ b_0 = k')$ may be calculated from the *trigonometrical recurrence formula*

$$\sin \theta_{n+1} = \tan^2 \tfrac{1}{2}\theta_n,$$

commencing with $\sin \theta_0 = k$, $\sin \theta_1 = (1 - k')/(1 + k')$, etc.

Hence derive the following formulae :

(i) $c_1 = \tfrac{1}{2}k\tan \tfrac{1}{2}\theta_0$, $c_2 = \tfrac{1}{4}k\tan \tfrac{1}{2}\theta_0 \tan \tfrac{1}{2}\theta_1$, $c_3 = \tfrac{1}{8}k\tan \tfrac{1}{2}\theta_0 \tan \tfrac{1}{2}\theta_1 \tan \tfrac{1}{2}\theta_2$,

$c_1{}^2 = \tfrac{1}{4}k^2 \sin \theta_1$, $c_2{}^2 = \tfrac{1}{16}k^2 \sin \theta_1 \sin \theta_2$, $c_3{}^2 = \tfrac{1}{64}k^2 \sin \theta_1 \sin \theta_2 \sin \theta_3$,

from which, in general, expressions for c_n and $c_n{}^2$ may be formed.

[Note that $c_n/a_n = \sin \theta_n$, $b_n/a_n = \cos \theta_n$.]

(ii) $\log (2K/\pi) = \sum_1^\infty \log (1 + \sin \theta_n) = 2\sum_1^\infty \log \sec \tfrac{1}{2}\theta_{n-1}$.

(iii) $\log \{\pi/(2Kk')\} = \tfrac{1}{2}\sum_1^\infty \log \sec \theta_{n-1}$.

(iv) $\tfrac{1}{2}\pi K'/K = -\tfrac{1}{2}\log q = \tfrac{1}{2}\log (4/\sin \theta_1) - \sum_1^\infty (\tfrac{1}{2})^n \log (1 + \sin \theta_{n+1})$.

[To derive (iv), make use of formula (61).]

10. Show that the A.G.M. scale $(a_0 = 1,\ b_0 = k')$ may be calculated from the *hyperbolic scale of decreasing amplitudes*

$$\tanh 2\chi_{n+1} = \tanh^2 \chi_n, \quad \text{or} \quad \chi_{n+1} = \tfrac{1}{4}\log \cosh 2\chi_n,$$

commencing with $\tanh 2\chi_0 = k$, or $\chi_0 = \tfrac{1}{4}\log \{(1 + k)/(1 - k)\}$.

Hence derive the following formulae :

(i) $c_1 = \tfrac{1}{2}k\tanh \chi_0$, $c_2 = \tfrac{1}{4}k\tanh \chi_0 \tanh \chi_1$, $c_3 = \tfrac{1}{8}k\tanh \chi_0 \tanh \chi_1 \tanh \chi_2$,

$c_1{}^2 = \tfrac{1}{4}k^2 \tanh 2\chi_1$, $c_2{}^2 = \tfrac{1}{16}k^2 \tanh 2\chi_1 \tanh 2\chi_2$,

$c_3{}^2 = \tfrac{1}{64}k^2 \tanh 2\chi_1 \tanh 2\chi_2 \tanh 2\chi_3$,

with similar expressions for c_n and $c_n{}^2$.

[Note that $b_n/a_n = \operatorname{sech} 2\chi_n$, $c_n/a_n = \tanh 2\chi_n$.]

(ii) $\log(2K/\pi) = \sum_{1}^{\infty} \log(1 + \tanh 2\chi_n)$.

(iii) $\log\{\pi/(2Kk')\} = 2\sum_{1}^{\infty} \chi_n$.

(iv) $\frac{1}{2}\pi K'/K = \log q^{-\frac{1}{2}} = \log(2\coth\chi_0) - \sum_{1}^{\infty} (\frac{1}{2})^n \log(1 + \tanh\chi_{n+1})$.

[To derive (iv) make use of formula (61).]

11. Show that the A.G.M. scale $(a_0 = 1, b_0 = k')$ may be calculated from the *hyperbolic scale of increasing amplitudes*

$$\tanh 2\psi_n = \tanh^2 \psi_{n+1}, \text{ or } \psi_n = \tfrac{1}{4}\log\cosh 2\psi_{n+1},$$

commencing with $\tanh 2\psi_0 = k'$, or $\psi_0 = \frac{1}{4}\log\{(1 + k')/(1 - k')\}$.
Hence derive the following formulae:

(i) $c_{n+1}/c_n = \frac{1}{2}e^{-2\psi_n}$, $b_n/a_n = \tanh 2\psi_n$, $c_n/a_n = \operatorname{sech} 2\psi_n$.

(ii) $\log(2K/\pi) = \sum_{1}^{\infty} \log(1 + \operatorname{sech} 2\psi_n)$.

(iii) $\log\{\pi/(2Kk')\} = \sum_{0}^{\infty} \log\coth\psi_{n+1}$.

(iv) $\frac{1}{2}\pi K'/K = \log q^{-\frac{1}{2}} = \psi_n/2^{n-1}$, as n tends to infinite values.

[To derive (iv) make use of formula (60).]

12. Writing $\xi_n = \frac{1}{2}\{1 + (1 - e^{-8\psi_n})^{\frac{1}{2}}\}$, show that the recurrence formula of Example 11 may be written

$$2\psi_{n+1} - 4\psi_n = \log 2 + \log \xi_n.$$

Hence, denoting $\beta = (1 - \sqrt{k'})/(1 + \sqrt{k'})$, derive the expansions

(i) $\log q^{-1} = \log(2/\beta) + \sum_{0}^{\infty} (\frac{1}{2})^{n+1} \log \xi_{n+1}$.

(ii) $\log q = \log(\frac{1}{2}\beta) + 2(\frac{1}{2}\beta)^4 + 13(\frac{1}{2}\beta)^8 + \frac{368}{3}(\frac{1}{2}\beta)^{12} + \dots$.

(iii) $q = \frac{1}{2}\beta + 2(\frac{1}{2}\beta)^5 + 15(\frac{1}{2}\beta)^9 + 150(\frac{1}{2}\beta)^{13} + \dots$.

(See also Example 24.)

13. Derive the following series for π from the formulae of Example 12,

$$\pi = \log x^{-1} - 2x^4 - 13x^8 - \tfrac{368}{3}x^{12} - \dots, \text{ where } x = \tfrac{1}{2}(2^{\frac{1}{4}} - 1)/(2^{\frac{1}{4}} + 1).$$

[Write $k = k' = 1/\sqrt{2}$, then $K = K'$ and $\log q^{-1} = \pi$. Note that

$$x^{-1} = 23\cdot14086, \quad \log x^{-1} = 3\cdot141602\dots, \quad 2x^4 = \cdot0000070,$$

so that two terms of the above series give the value of π correctly to the tenth decimal. The remarkable approximation represented by the first term of the series was first pointed out by Legendre, *F. E.* t. L.]

14. Given the *nome* q, compute the moduli k, k'.

[The recurrence formula of Example 11 may be written

$$4\psi_n - 2\psi_{n+1} = -\log 2 + \log(1 - e^{-4\psi_{n+1}})$$
$$= -\log 2 - e^{-4\psi_{n+1}} - \tfrac{1}{2}e^{-8\psi_{n+1}} - \dots$$

Note that when n is sufficiently large $(\tfrac{1}{2})^{n+1}\psi_{n+1}=\tfrac{1}{4}\log q^{-1}$. We then have

$$4\psi_n = 2^n \log q^{-1} - \log 2 - q^{2^{n+1}}.$$

Choose n so that the last term is negligible compared to the first two. The repeated application of the above recurrence formula enables us to calculate successively $\psi_{n-1} \dots \psi_2$ and finally ψ_0, in terms of which $k' = \tanh 2\psi_0$, $k = \operatorname{sech} 2\psi_0$.]

15. Derive the differential relations

$$\frac{1}{2^n}\frac{1}{a_n{}^2} d\left(\log\frac{c_n}{b_n}\right) = \frac{1}{2^n}\frac{1}{b_n{}^2} d\left(\log\frac{c_n}{a_n}\right) = \frac{1}{2^n}\frac{1}{c_n{}^2} d\left(\log\frac{a_n}{b_n}\right) = \frac{1}{c_0{}^2} d\left(\log\frac{a_0}{b_0}\right) = \nabla$$

for all values of n.   Gauss, *G. W.* Bd. III. p. 380.

16. From the results of Example 15 obtain directly the formula

$$\operatorname*{Lt}_{n\to\infty}\left(\frac{1}{2^n}\log\frac{a_n}{c_n}\right) = \tfrac{1}{2}\pi\frac{a_n}{a_n'}.$$

[In terms of the A.G.M. scale (a_0, b_0, c_0) and the complementary scale (a_0', b_0', c_0'), the formulae of Example 15 give

$$\frac{1}{2^n}\frac{1}{a_n{}^2} d\left(\log\frac{c_n}{b_n}\right) = \frac{1}{c_0{}^2} d\left(\log\frac{a_0}{b_0}\right) \text{ and } \frac{1}{2^n}\frac{1}{a_n'^2} d\left(\log\frac{c_n'}{b_n'}\right) = \frac{1}{c_0'^2} d\left(\log\frac{a_0'}{b_0'}\right).$$

But $a_0 = a_0'$, $b_0 = c_0'$, $c_0 = b_0'$ and $a_0{}^2 = b_0{}^2 + c_0{}^2$, from which it is easily proved that

$$\frac{1}{c_0{}^2} d\left(\log\frac{a_0}{b_0}\right) + \frac{1}{c_0'^2} d\left(\log\frac{a_0'}{b_0'}\right) = 0.$$

Hence we conclude that $\dfrac{1}{a_n{}^2} d\left(\dfrac{1}{2^n}\log\dfrac{c_n}{b_n}\right) + \dfrac{1}{a_n'^2} d\left(\dfrac{1}{2^n}\log\dfrac{c_n'}{b_n'}\right) = 0.$

We also have the identity $\dfrac{1}{a_n{}^2} d\left(\dfrac{a_n}{a_n'}\right) + \dfrac{1}{a_n'^2} d\left(\dfrac{a_n'}{a_n}\right) = 0$, from which it

follows that $\dfrac{1}{2^n}\log\dfrac{c_n}{b_n} = C_n\dfrac{a_n}{a_n'}$, C_n being a constant independent of

(a_n, b_n, c_n), (a_n', b_n', c_n'). In terms of Gauss' notation for the A.G.M., we note from Section III that

$$a_n' = M(a_0, c_0) = 2^n M(a_n, c_n) = 2^n a_n M(1, c_n/a_n),$$

while $a_n = M(a_0, b_0) = M(a_n, b_n)$. Thus when n becomes indefinitely

great, $C_n = \dfrac{a_n'}{a_n}\dfrac{1}{2^n}\log\dfrac{c_n}{b_n} = \dfrac{1}{M(1, c_n/a_n)}\log\dfrac{c_n}{a_n}$. But if we write $k' = c_n/a_n$,

$M(1, c_n/a_n) = \tfrac{1}{2}\pi/K = \tfrac{1}{2}\pi/\log(4/k')$ as k' becomes small. Thus, in the limit as n increases indefinitely, $C_n = -\tfrac{1}{2}\pi$, so that

$$\operatorname*{Lt}_{n\to\infty}\left(\frac{1}{2^n}\log\frac{a_n}{c_n}\right) = \tfrac{1}{2}\pi\frac{a_n}{a_n'}.]$$

17. Making use of the result of Example 16, prove that each of the ratios in Example 15 is given by

$$\nabla = \operatorname*{Lt}_{n \to \infty} \; [\tfrac{1}{2}\pi/(a_n a_n') . \, d \log (a_n'/a_n)].$$

18. Derive Legendre's relation $E/K + E'/K' - 1 = \tfrac{1}{2}\pi/(KK')$ directly from the properties of the A.G.M. scales $(a_0 = 1,\ b_0 = k')$ and $(a_0' = 1,\ b_0' = k)$.

[Differentiating the identity $\log (a_n b_n) - \log (a_{n-1} b_{n-1}) = \log (a_n/b_n)$ and making use of the results of Example 15, we have

$$d \log (a_n b_n) - d \log (a_{n-1} b_{n-1}) = d \log (a_n/b_n) = 2^n \, (c_n^2/c_0^2) \, d \log (a_0/b_0).$$

Writing down similar equations for $n-1$, $n-2$, ... 3, 2, 1 in place of n and adding, we find

$$d \log (a_n b_n) - d \log (a_0 b_0) = (2c_1^2 + 4c_2^2 + ... + 2^n c_n^2) \, c_0^{-2} \, d \log (a_0/b_0) \quad \text{(i)}$$

and

$$d \log (a_n' b_n') - d \log (a_0' b_0') = (2c_1'^2 + 4c_2'^2 + ... + 2^n c_n'^2) \, c_0'^{-2} \, d \log (a_0'/b_0') \\ \qquad\qquad ......\text{(ii)}.$$

Since $a_0 = a_0' = 1$ and $b_0' = c_0$, we have

$$d \log (a_0' b_0') - d \log (a_0 b_0) = d \log (c_0/b_0) = a_0^2 c_0^{-2} \, d \log (a_0/b_0).$$
$$\text{[Ex. 15 with } n = 0.]$$

When n becomes large, $a_n = b_n$, so that

$$d \log (a_n b_n) - d \log (a_n' b_n') = 2d \log (a_n/a_n') = - (4/\pi) \, a_n a_n' c_0^{-2} \, d \log (a_0/b_0).$$
$$\text{[Ex. 17.]}$$

Note also the identity $c_0^{-2} \, d \log (a_0/b_0) + c_0'^{-2} \, d \log (a_0'/b_0') = 0.$ [Ex. 16.]

Subtracting (ii) from (i) and dividing out by

$$c_0^{-2} \, d \log (a_0/b_0) = - c_0'^{-2} \, d \log (a_0'/b_0')$$

we find

$$(4/\pi) \, a_n a_n' - a_0^2$$
$$= - (2c_1^2 + 4c_2^2 + ... + 2^n c_n^2 + ...) - (2c_1'^2 + 4c_2'^2 + ... + 2^n c_n'^2 + ...)$$
$$\qquad\qquad ......\text{(iii)}.$$

But we have $(K - E)/K = \tfrac{1}{2} (c_0^2 + 2c_1^2 + ... + 2^n c_n^2 + ...)$, $a_n = \tfrac{1}{2}\pi/K$, $a_0 = 1$ with similar expressions for the complementary functions in terms of accented symbols. Thus (iii) becomes

$$\pi/(KK') - 1 = - 2 (1 - E/K) + c_0^2 - 2 (1 - E'/K') + c_0'^2$$

or noting that $c_0^2 + c_0'^2 = 1$, we derive Legendre's relation in the familiar form

$$E/K + E'/K' - 1 = \tfrac{1}{2}\pi/(KK').$$

The equivalent of the series (iii) is given by Gauss (G. W. Bd. III. p. 380), but it does not seem to have been hitherto pointed out that Legendre's relation follows immediately from it.]

19. Verify the following identities:

$$(1 + 2x + 2x^4 + 2x^9 + \ldots)(1 - 2x + 2x^4 - 2x^9 + \ldots)$$
$$= (1 - 2x^2 + 2x^8 - 2x^{18} + \ldots)^2,$$

$$(1 + 2x + 2x^4 + 2x^9 + \ldots)^2 + (1 - 2x + 2x^4 - 2x^9 + \ldots)^2$$
$$= 2(1 + 2x^2 + 2x^8 + 2x^{18} + \ldots)^2,$$

$$(1 + 2x + 2x^4 + 2x^9 + \ldots)^4 - (1 - 2x + 2x^4 - 2x^9 + \ldots)^4$$
$$= (2x^{\frac{1}{4}} + 2x^{\frac{9}{4}} + 2x^{\frac{25}{4}} + \ldots)^4.$$

Gauss, Nachlass, G. W. Bd. III. p. 447.

20. From the results of Example 19, prove that if we form the A.G.M. scale commencing with

$$a_0 = h(1 + 2x + 2x^4 + \ldots)^2, \qquad b_0 = h(1 - 2x + 2x^4 - \ldots)^2,$$
$$c_0 = h(2x^{\frac{1}{4}} + 2x^{\frac{9}{4}} + 2x^{\frac{25}{4}} + \ldots)^2,$$

the nth term is given by

$$a_n = h(1 + 2x_n + 2x_n^4 + \ldots)^2, \qquad b_n = h(1 - 2x_n + 2x_n^4 - \ldots)^2,$$
$$c_n = h(2x_n^{\frac{1}{4}} + 2x_n^{\frac{9}{4}} + 2x_n^{\frac{25}{4}} + \ldots)^2,$$

where $\log x_n = 2^n \log x$ and $h = M(a_0, b_0)$.

Gauss, Nachlass, G. W. Bd. III. p. 448.

21. Prove that x as defined in Example 19 is identical with the nome q when $(a_0 = 1, b_0 = k')$.

[When n becomes large,

$$-\tfrac{1}{2}\log q = \tfrac{1}{2}\pi K'/K = \mathop{\mathrm{Lt}}_{n \to \infty} \frac{1}{2^n} \log(a_n/c_n)$$
$$= \mathop{\mathrm{Lt}}_{n \to \infty} \frac{1}{2^n} \log[(1 - 2x_n + \ldots)^2/4x_n^{\frac{1}{2}}(1 + 2x_n^2 + \ldots)^2]$$
$$= -\mathop{\mathrm{Lt}}_{n \to \infty} \frac{1}{2^{n+1}} \log x_n$$
$$= -\tfrac{1}{2}\log x,$$

so that $x = q$.]

22. From the results of Examples 20 and 21, derive the following expressions:

$$\sqrt{k'} = \frac{1 - 2q + 2q^4 - 2q^9 + \ldots}{1 + 2q + 2q^4 + 2q^9 + \ldots}, \qquad \sqrt{k} = \frac{2q^{\frac{1}{4}} + 2q^{\frac{9}{4}} + 2q^{\frac{25}{4}} + \ldots}{1 + 2q + 2q^4 + 2q^9 + \ldots},$$

$$K = \tfrac{1}{2}\pi(1 + 2q + 2q^4 + \ldots)^2, \qquad Kk' = \tfrac{1}{2}\pi(1 - 2q + 2q^4 - \ldots)^2.$$

[From Example 20 we obviously have $a_0 = 1$, $b_0 = k'$, $c_0 = k$, $h = \tfrac{1}{2}\pi/K$.]

23. If we denote $\beta = (1 - \sqrt{k'})/(1 + \sqrt{k'})$, derive the following expressions:

$$\tfrac{1}{2}\beta = (q + q^9 + q^{25} + \ldots)/(1 + 2q^4 + 2q^{16} + \ldots),$$
$$K = 2\pi(1 + 2q^4 + 2q^{16} + \ldots)^2/(1 + \sqrt{k'})^2 = 8\pi(q + q^9 + q^{25} + \ldots)^2/(1 - \sqrt{k'})^2.$$

24. From the preceding examples derive the well-known expansion
$q = \frac{1}{2}\beta + 2\left(\frac{1}{2}\beta\right)^5 + 15\left(\frac{1}{2}\beta\right)^9 + 150\left(\frac{1}{2}\beta\right)^{13} + \ldots$ etc. (See also Example 12.)

25. If we write
$$a_n = 1 + 2x_n + 2x_n^4 + \ldots, \qquad \beta_n = 1 - 2x_n + 2x_n^4 - \ldots,$$
$$\gamma_n = 2x_n^{\frac{1}{4}} + 2x_n^{\frac{9}{4}} + 2x_n^{\frac{25}{4}} + \ldots,$$
where $\log x_n = 2^n \log q$, prove that the recurrence formula
$$\tan \psi_{n+1} = (a_{n+1}/\beta_{n+1}) \tan 2\psi_n$$
leads to the result
$$\underset{n \to \infty}{\mathrm{Lt}} \ (\psi_n/2^n) = (\tfrac{1}{2}\pi/K)\, F(\phi, k),$$
where $\tan \psi_0 = \sqrt{k'} \tan \phi$, and K is given by the q-series of Example 22.

26. In terms of Landen's scale of increasing amplitudes (Section IV)
$\tan (\phi_{n+1} - \phi_n) = (b_n/a_n) \tan \phi_n$, derive the following expansions :
$$\mathrm{dn}\,(u, k) = \Delta\,(\phi_0, k) = a_n + c_1 \cos \phi_1 + c_2 \cos \phi_2 + \ldots + c_n \cos \phi_n + \ldots,$$
$$k^2 \,\mathrm{sn}\,u\,\mathrm{cn}\,u = \tfrac{1}{2} \sum_0^\infty 2^{n-1} c^2_{n-1} \sin \phi_n \cos (2\phi_{n-1} - \phi_n).$$

27. When (c_{n+1}/a_{n+1}) is sufficiently small, prove that ϕ_{n+1} of Example 26 may be computed from the expansion
$$\phi_{n+1} = 2\phi_n - (c_{n+1}/a_{n+1}) \sin 2\phi_n$$
$$+ \tfrac{1}{2}(c_{n+1}/a_{n+1})^2 \sin 4\phi_n - \tfrac{1}{3}(c_{n+1}/a_{n+1})^3 \sin 6\phi_n - \ldots.$$

28. Prove that the recurrence formula of Example 26 is equivalent to $\Delta_n = \tfrac{1}{2}(\Delta^2_{n-1} + b^2_n)/\Delta_{n-1}$, and hence show that
$$\log [\Theta\,(u)/\Theta\,(0)] = \sum_0^\infty \frac{1}{2^{n+1}} \log (a_n/\Delta_n).$$

[Jacobi, *Fundamenta Nova* (1881), § 52, p. 204.]

29. Prove that the recurrence formula $a_n \tan \phi_{n+1} = \Delta_{n+1} \tan \phi_n$, where $\Delta_n = \sqrt{(a_n^2 \cos^2 \phi_n + b_n^2 \sin^2 \phi_n)}$, may also be written in the forms
$$\frac{\sin \phi_{n+1}}{\sin \phi_n} = \frac{2a_{n+1}}{a_n + \Delta_n}, \quad \frac{\tan \phi_{n+1}}{\tan \phi_n} = \frac{\sqrt{a_{n+1}}}{\sqrt{a_n}}\frac{\sqrt{b_n + \Delta_n}}{\sqrt{a_n + \Delta_n}}, \quad \frac{\Delta_{n+1}}{\sqrt{a_n a_{n+1}}} = \frac{\sqrt{b_n + \Delta_n}}{\sqrt{a_n + \Delta_n}}.$$

30. Prove that as a result of the recurrence formula of Example 29, $d\phi_n/\Delta_n = d\phi_{n+1}/\Delta_{n+1}$. Hence derive the following formulae :
$$F(\phi, k) = \underset{n \to \infty}{\mathrm{Lt}}\ (\phi_n/a_n), \text{ where } \phi = \phi_0, \ \sin \phi_0 = \mathrm{sn}\,(u, k),$$
$$Z\,(\phi, k) = 2c_1 \cos \phi_0 \sin \phi_1 + 4c_2 \cos \phi_1 \sin \phi_2 + \ldots + 2^n c_n \cos \phi_{n-1} \sin \phi_n + \ldots,$$
$$\log [\Theta\,(u, k)/\Theta\,(0, k)] = \sum_0^\infty 2^n \log [2a_n/(a_n + \Delta_n)].$$

[The recurrence formula which lies at the basis of the results obtained in Examples 29 and 30 was originally given by Gauss in 1818 in his paper "Determinatio Attractionis, etc." *G. W.* Bd. III. (1866), p. 353, and was further discussed by Jacobi, "Zur Numerischen Berechnung der Elliptischen Functionen," *G. W.* Bd. I. (1881), p. 351. It is easily seen that these formulae are not adapted to rapid numerical calculation.]

31. If $\sin \phi_0 = \operatorname{sn}(u, k)$ prove that

$$\tan a_n u = \frac{\Delta_1 \Delta_2 \ldots \Delta_n \ldots}{a_0 a_1 a_2 \ldots a_n \ldots} \tan \phi_0.$$

[The result follows from Gauss' recurrence formula of Example 29 ; see also Jacobi, *F. N.*, §36, *G. W.* Bd. I. p. 154.]

32. Prove that Landen's recurrence formula

$$\tan (\psi_n - \psi_{n+1}) = (c_{n+1}/a_{n+1}) \tan \psi_{n+1}$$

may be written $\tan 2\chi_{n+1} = \sqrt{(c_n/a_n)} \tan \chi_n$ if we write

$$\tan \psi_n = \sqrt{(a_n/c_n)} \tan \chi_n.$$

33. If we make use of Landen's scale of decreasing amplitudes

$$\sin (2\psi_{n+1} - \psi_n) = (b_n/a_n) \sin \psi_n,$$

commencing with $\psi_0 = \frac{1}{2}\pi$, prove that $K' = (2K/\pi) \log \tan (\frac{1}{4}\pi + \frac{1}{2}\psi_n)$ when n increases indefinitely. Hence show that the nome q may be computed from the formula $q = \cot^2 (\frac{1}{4}\pi + \frac{1}{2}\psi_n) = (1 - \sin \psi_n)/(1 + \sin \psi_n)$. Show, furthermore, that Legendre's relation leads to the identity

$$a_n (1 - \sin \psi_n) = (\tfrac{1}{2}\pi/K) \{1 - \tanh (\pi K'/2K)\} = \underset{2}{\Sigma} (\psi_n, c_n),$$

where $\underset{2}{\Sigma} (\psi_n, c_n)$ is the series of equation (75).

34. Prove that if the A.G.M. scale (a_0, b_0) is computed, the circumference l of an ellipse of semi-axes (a_0, b_0) is given by

$$l = (2\pi/a_n) (a_1^2 - 2c_2^2 - 4c_3^2 - 8c_4^2 - \ldots).$$

35. If we denote by r_1 and r_2 the greatest and least distances respectively of a point on one circle to the circumference of a coaxial one, show that in terms of the A.G.M. scale $(a_0 = r_1, b_0 = r_2)$ the coefficient of mutual induction is given by

$$M = \{2\pi^2/a_n\} [c_1^2 + 2c_2^2 + 4c_3^2 + \ldots + 2^{n-1}c_n^2 + \ldots],$$

or, when the circles are very close together, in terms of the complementary scale $(a_0' = r_1 + r_2, b_0' = r_1 - r_2)$, by the formula

$$M = 2\pi a_n' \left[\frac{(r_1^2 - 2c_2'^2 - 4c_3'^2 - \ldots)}{a_n'^2} \left(\log \frac{4a_2'}{r_2} - \tfrac{1}{2} \log \frac{a_2'}{a_3'} - \tfrac{1}{4} \log \frac{a_3'}{a_4'} - \ldots \right) - 2 \right].$$

[King, L. V., *Proc. Roy. Soc.* A, Vol. C. 1921, p. 63 *et seq.* Several numerical illustrations are given in this paper.]